The Early Use of Iron in India

The Early Use of Iron in India

DILIP K. CHAKRABARTI

DELHI
OXFORD UNIVERSITY PRESS
BOMBAY CALCUTTA MADRAS
1992

Oxford University Press, Walton Street, Oxford OX2 6DP
New York Toronto
Delhi Bombay Calcutta Madras Karachi
Kuala Lumpur Singapore Hong Kong Tokyo
Nairobi Dar es Salaam
Melbourne Auckland
and associates in
Berlin Ibadan

SBN 0 19 562992 2

Typeset by Imprinter, C-79, Okhla Ind. Area, Phase-I, New Delhi 110020
Printed at Rekha Printers (P) Ltd., New Delhi 110020
and published by S. K. Mookerjee, Oxford University Press
YMCA Library Building, Jai Singh Road, New Delhi 110001

For Frank Raymond Allchin

Contents

CHAPTER FIVE
The Pre-Industrial Iron-Smelting Tradition

CHAPTER SIX
Summary and Conclusions

List of Illustrations

Preface

This work sets out to analyse the data on the early use of iron in India. This is not a book about the Indian Iron Age. Its emphasis is on the occurrence of iron artifacts at some representative sites in different regions and the metallurgical studies of these artifacts where such studies are available. At the same time, the geological data on the distribution of iron ores, the literary references to the different aspects of the use of iron and other issues and, finally, the ethnographic accounts of pre-industrial iron-smelting practices in some parts of the country have been studied to lend the correct perspective to this core of archaeological and related metallurgical evidence. We realize that a proper breakthrough in the study of the use of different metals in India can come only through a wide and co-ordinated metallurgical study of the relevant specimens from different periods and regions. Meanwhile, this volume is intended to be a definitive (although not exhaustive) study of the sources on the early use of iron in India. Methodologically, this is rooted in 'Panchanan Neogi's *Iron in Ancient India* published as early as 1914. Neogi was a chemist who had the sagacity to realize that the study of ancient Indian iron (or metals in general) depended on geological, archaeological, metallurgical, literary and ethnographic sources.

This study has been vitiated in recent years by a number of ill-founded notions: (a) Iron was a great harbinger of socio-economic changes in the context of early historic India. (b) Iron was related to the arrival of the Aryans or Indo-Iranian language-speaking groups in the country.

There is also a great reluctance to admit the occurrence of iron in the Indian archaeological context in the second half of the second millennium B.C. The arguments that iron at Aḥar comes from pits or from trenches laid on the slope of the mound ignore the fact that iron objects and slag have been reported not merely from the uppermost 'chalcolithic' level, namely, Ahar IC, but from the earlier IB as well. It is interesting to observe that in a paper on the protohistoric West Bengal which was presented at the seminar, 'Archaeology in Eastern India', held in February 1989 in Patna, Sudhin De of the West Bengal State Directorate of Archaeology and Museums reported the occurrence of 'iron slags and other artifacts from the stratified chalcolithic level at Pandu Rajar Dhibi (1985) and Mangalkot (1988) in Burdwan district'. In her article 'Archaeology of Mangalkot' published in A. Ray and S. Mukherjee (eds.), *Historical Archaeology of India*, Delhi, 1990, pp. 131–40 Amita Ray points out, on the basis of the occurrence of iron slags, ingots and iron tools like the arrowhead, point and spearhead from the lowest strata of Mangalkot, that 'the chalcolithic people in this region had used iron from the very beginning as a form of technology'. This discovery clearly substantiates my earlier (1981) find at Bahiri in the Birbhum district in the same region.

In the context of Kashmir, a very significant discovery has recently been announced by A. K. Sharma who, in his article 'Neolithic Gufkral' published in *Indian Archaeological Heritage* (eds. C. Margabandhu, K. S. Ramachandran, A. P. Sagar, D. K. Sinha), Delhi, 1991, pp. 101–10, writes, 'the most important point is the date of iron at Gufkral which as per C-14 dating comes to around 1400 B.C.' Discoveries such as these strongly support, among other things, the idea of the existence of iron at Ahar as early as the middle of the second millennium B.C.

It is also important to note that the principle of iron-working developing out of the copper-smelting tradition has found a general acceptance among the metallurgists working in this field. For instance, in their paper 'Reflections on Early Metallurgy in Southeast Asia' in *The Beginning of the Use of Metals and Alloys* (ed. Robert Maddin, 1988, Boston), Tamara Stech and Robert Maddin quite clearly accept the notion that 'iron could have been encountered first under certain conditions in the copper-smelting furnace, if iron oxide contained in copper ore (for example chalcopyrite) or a haematite flux was used for smelting the siliceous ore . . . There is at present no reason to suppose that iron-smelting was not an indigenous southeast Asian development'.

Finally, it may be fitting to draw attention to something which has possibly agitated the mind of every researcher of ancient iron and which has been rather well expressed by G. L. Possehl in his draft paper 'The Coming of the South Asian Iron Age: An Indigenous

Technological Process' (October 1988): 'The seemingly coincidental nature of the coming of the Iron Age near the end of the second millennium B.C. from the eastern Mediterranean across Asia to the South China Sea presents us with a need to measure the implications of this and consider common historical processes.'

The basic work on the book was done between October 1974 and March 1975 at the Institute for Advanced Studies in the Humanities, Edinburgh University, with the support of Professor A. M. Snodgrass, and in Wolfson College, Cambridge in 1975–6 with the support of Dr F. R. Allchin. Several papers were published on the basis of this work. However, the topic needed a full-length study, and I am glad that it has eventually become possible to complete it.

This was written during my stay in Jahangirnagar University, Bangladesh, and Nayanjot Lahiri kindly sent me xeroxed pages of books, etc. from time to time. I must also acknowledge a grant of Rs 1000 from the Indian Council of Historical Research for this work.

Dilip K. Chakrabarti

CHAPTER ONE

Research on Early Indian Iron

The history of research on Indian pre-industrial iron, the origin and antiquity of iron in India and the Indian Iron Age is well defined and goes back to the eighteenth century; it may be studied with reference to four phases leading up to the present.[1]

THE FIRST PHASE (c.1795–c.1850)

The primary feature of this phase was a scientific interest in understanding the properties and the manufacturing process of pre-industrial Indian steel which came to be known in Europe as *wootz*. This word was derived from the Canarese term *ukku* meaning steel.[2] That pre-industrial Indian steel had a reputation in contemporary Europe is exemplified by the following two records. First, it has been mentioned, presumably in the eighteenth-century context, that the Westphalians used to roll puddled steel into half-inch or three-fourth inch squares and sell them in Hamburg under the name of Indian steel.[3] Second, in 1722, the French scholar R. A. F. de Reaumur wrote in his memoir, 'On Methods of Recognizing Defects and Good Quality in Steel and on Several Ways of Comparing Different Grades of Steel' that, on the order of the Duke of Orleans, the French consul in Cairo,

[1] Chakrabarti, 1977.

[2] Yule and Burnell, 1903: 972–3.

[3] Lester, 1912: 4. The following comment of Lester (1912: 20) is a little nostalgic: 'the man who hopes to establish an iron or steel industry in India must not remain in the cities only, but must get out among the clever native craftsmen—the men who had been, the men who to some extent were, but never would be again, so far as steel and iron went in India.'

M. le Maire, sent him specimens of steel out of which the Damascus swords were supposed to have been made. 'Among the steels which I received from him, and which he assured me were the best, there is a cake which is supposed to be steel from India and the kind to be rated most highly in Egypt.' Reaumur does not seem to have carried out any specific research on this steel but he mentions that he could find 'no artisan in Paris who succeeded in forging a tool out of it' and that 'it withstood the fire hardly better than cast iron'.[4]

The first significant publication on the subject was by George Pearson in 1795. Pearson, a Fellow of the Royal Society, published in the *Philosophical Transactions of the Royal Society of London* his paper, 'Experiments and Observations to Investigate the Nature of a Kind of Steel, Manufactured at Bombay and there called *Wootz*: with Remarks on the Properties and Composition of the Different States of Iron'.[5] The specimens of wootz were in the shape of cakes, each of which was five inches in diameter and one inch thick and weighed more than two pounds. They were sent by Dr Helenus Scott of Bombay to Sir Joseph Banks, the President of the Royal Society. Pearson's theory that this steel was made directly from the ore and that it was never in the state of wrought iron was subsequently proved wrong. And so also was his suggested derivation of the term, wootz, from the word for steel in the Gujarati language spoken in Bombay. David Mushet pointed out soon afterwards, both in the light of his own research and the records of travellers, that wootz steel was never formed directly from the ore but by the fusion of fragments of small bars of malleable iron in a crucible with woody carbonaceous matter covered with leaves and clay. He felt that the term itself, wootz, was more likely to have been derived from the Canarese word, ukku.[6]

[4] Sisco, 1956: 176.

[5] Pearson, 1795: 322–46. Pearson (1795: 345) wrote: 'Although no account is given by Dr Scott of the process for making wootz, we may without much risk conclude that it is made directly from the ore; and consequently that it has never been in the state of wrought iron.'

[6] Mushet, 1805: 163–75. In 1840 Mushet (1840: 662) wrote: 'Notwithstanding the many imperfections with which wootz is loaded, it certainly possesses the radical principles of good steel, and impresses us with a high opinion of the ore from which it is formed. The possession of this ore for the fabrication of steel and bar iron might to this country be an object of the highest importance. At present it is a subject of regret that such a source of wealth cannot be annexed to its capital and talent.' He (1840: 664) further wrote: 'Wootz steel is never formed directly from the ore, but by the fusion of fragments of small bars of malleable iron, in a crucible with woody carbonaceous matter, as practised by myself long before a knowledge of this fact—which was so late as in the year 1825—first reached me.'

However, Pearson's paper is significant from the historical point of view, and along with the earlier interest shown by Reaumur in France, marks the beginning of scientific metallurgical interest in pre-industrial Indian iron.

In 1818, James Stodart wrote about wootz and 'its fitness for making surgical instruments, and other articles of fine cutlery' arguing that, when properly treated, wootz was 'vastly superior to the best cast-steel of Europe'.[7] In 1819, Michael Faraday tried to determine if wootz contained any substances other than iron and carbon.[8] Wootz was also analysed by J. R. Breant, Inspector of Assays at the Paris Mint, in 1823.[9] Robert Hadfield has explained why wootz drew such a wide experimental interest in this period: '*Wootz* . . . was in many respects superior to anything that the steel-makers of Western Europe had hitherto produced. The investigation of its properties and the attempt to imitate them was a very commendable research.'[10]

Soon afterwards there were attempts to determine the cause of the external pattern of the famous Damascus sword-blades. In 1836, Henry Wilkinson showed that the steel used for these blades was Indian steel and that the pattern on them depended on the crystallized nature of the cakes of Indian steel. It is worthwhile to cite Wilkinson at some length on this point:

> This is the kind of steel (i.e. wootz) that has always been employed for these blades so celebrated for their beauty, and I consider that it would be as impossible to forge a sword-blade out of some of these materials, when properly selected, without obtaining the true Damascus figure as it would be to imitate the pattern by any contortions of iron or steel artificially. These cakes of steel are evidently crystallized, and the future pattern of the sword-blades depends on the size and arrangement of the crystals, modified by so many circumstances, that it is not surprising the proper kind of steel for this purpose should be so rare, or that the secret should

[7] Stodart, 1818: 570–1. 'Wootz, when treated in the manner that has been pointed out, is fit to be manufactured into cutlery of every description; it is invaluable for surgical instruments, where mediocrity is not, at least ought not to be admitted, as the sufferings of the patient are much lessened in intensity and duration, when fortunately the operator is freed from embarassments necessarily connected with a bad or indifferent instrument, and provided with one, the forms and physical properties of which are exactly to his wish.'

[8] Faraday, 1819.

[9] Breant, 1823. Breant conducted a series of 300 experiments using for this purpose 100 kg of wootz given to him by the East India Company.

[10] Hadfield, 1931: 67.

> have been supposed to be lost . . . it may be rendered self-evident that the figure or pattern, so long sought after, exists in the cakes of *wootz* or native steel of India, and only requires to be produced by the action of diluted acids.[11]

In 1839 Wilkinson requested the members of the Royal Asiatic Society to send him specimens of iron ore, crucibles and of wootz with a few specific details which he listed.[12] About this time, Colonel Anossoff of the Corps of Engineers, Imperial Army of Russia, and Master of the Fabric of Arms at Zlatoost, revived the Damask pattern.[13] In 1847 James Abbott, who was interested in Anossoff's work, described the 'process of working the Damascus blade of Goojrat (Punjab)'.[14]

The next important feature of research in this phase was the accumulation of data on the process of manufacture of wootz and ordinary iron as obtained during that period in different parts of India. As far as wootz was concerned, the most significant reports were by Francis Buchanan, Benjamin Heyne and E. H. Voysey, all of whom set down their observations in the context of south India, particularly Karnataka, Kerala and Andhra Pradesh. Francis Buchanan's *A Journey from Madras through the Countries of Mysore, Canara and Malabar* was published in three volumes in 1807.[15] Benjamin Heyne's *Tracts, Historical and Statistical on India* was published in 1814.[16] Buchanan's journey was undertaken on the order of Lord Wellesley, who was then the Governor-General of the British territories in India, 'for the express purpose of investigating the state of agriculture, arts and commerce, the religion, manners and customs, the history, natural and civil, and antiquities in the dominions of Mysore and the countries acquired by the Honourable East India Company . . . from Tipoo Sultan'. It is not known whether Heyne was similarly deputed; in the title page of his book he is introduced as the 'surgeon and naturalist on the establishment of Fort St. George', Madras. Buchanan's description of iron and steel manufacture came from a few places in Coimbatore, Malabar and Mysore; and Heyne discussed it in the context of Mysore and the southern part of modern Andhra Pradesh. An additional feature of interest in Voysey's report is that he saw traders from Isfahan at

[11] Wilkinson, 1836: 190.
[12] Wilkinson, 1839: 383–9.
[13] In Abbott, 1843, vol. II: lxx-lxxxviii.
[14] Abbott, 1847.
[15] Buchanan, 1807.
[16] Heyne, 1814.

Kona Samudram on the Godavari in Andhra Pradesh, who had come to purchase local steel.[17]

An account of iron-smelting practised by the Kol tribe of eastern India found a place in Buchanan's survey of the district of Bhagalpur in 1810–11 but that was not published till 1930.[18] The other reports were not long in coming. They are significantly large in number and one can mention only a few of them: an anonymous account of iron-smelting in Ferozepur (Punjab) in 1831,[19] Robert Rose's 'Account of the Process of Making Iron at Amdeah near Sambalpur (Orissa)' in 1831,[20] W. Cracroft's description of 'Smelting of Iron in the Kasya (Khasia) Hills (Assam)' in 1832,[21] J. M. Heath's account of the south Indian process in 1832[22] and 1839,[23] Legrand Jacob's report on Kathiawar in 1843,[24] Lt. Yule's report on the Khasia hills in 1842,[25] M. Marcadieu's report on Kulu in 1855,[26] Bal Gangadhar Shastree's report on Malvan in Ratnagiri district of Maharashtra in 1844,[27] S. F. Hannay's report on upper Assam in 1857,[28] James Abbott's account of gun-manufacture at Koteli near Sialkot (Punjab) in 1848,[29] the reports of Henry Piddington, Welby Jackson, H. Torrens and W. E. Baker on Birbhum and Ranigunj (West Bengal) in 1855,[30] 1845,[31] 1851[32] and 1854,[33] and finally, T. H. Newbold's examination of the nature, distribution and methods in 1846.[34] This list of publications is by no means exhaustive but, taken together, these writings made clear the extent, distribution and methods of pre-industrial iron-smelting in India. Geology was not emphasized, nor was any particular emphasis laid on the determination of its antiquity

[17] Voysey, 1832.
[18] Oldham, 1930.
[19] Anonymous, 1831.
[20] Rose, 1831.
[21] Cracroft, 1832.
[22] Heath, 1832.
[23] Heath, 1839.
[24] Jacob, 1843.
[25] Lt Yule, 1842.
[26] Marcadieu, 1855.
[27] Shastree, 1844.
[28] Hannay, 1857.
[29] Abbott, 1848.
[30] Piddington, 1855.
[31] Jackson, 1845.
[32] Torrens, 1851.
[33] Baker, 1854.
[34] Newbold, 1846.

through literature and archaeology. Incidentally, it is worth recalling that Heyne included in his volume a letter from J. Stodart, 'an eminent instrument-maker' who wrote on the basis of his 'own practice and experience' that wootz or steel of India was 'decidedly the best' he had met with and that it promised to be of importance to the manufacturers of Britain.

There was only a marginal interest in the origin and antiquity of Indian iron in this phase of research but, apparently, there was no doubt about India being the original centre of steel manufacture in the ancient context. J.M. Heath was one of the directors of the Indian Iron and Steel Company which was set up in 1833 and had its base of operations in the south. On the problems of antiquity and origin, J.M. Heath seems to have been the only writer of this period who had an interest and an opinion:

> The antiquity of the Indian process is no less astonishing than its ingenuity. One can hardly doubt that the tools with which the Egyptians covered their obelisks and temples of porphyry and syenite with hieroglyphs were made of Indian steel. There is no evidence to show that any of the nations of antiquity besides the Hindus were acquainted with the art of making steel.[35]

In this context he referred to the Classical writers, particularly to Quintus Curtius, who mentioned an Indian present of steel to Alexander. Of its origin in India, Heath apparently had no doubt: 'It appears then that the claim of India to a discovery which has exercised more influence on the arts conducive to civilization and manufacturing industry, than any other within the whole range of human inventions, is altogether unquestionable.'[36]

The attraction which wootz exercised during this period can be amply gauged by J. Stodart's trade card in London around 1820 which read, 'J. Stodart . . ., surgeon's instruments, razors and other cutlery made from . . . a steel from India, preferred by Mr Stodart to the best steel in Europe after years of comparative trial.'[37]

THE SECOND PHASE (c.1850–c.1950)

There are at least six interrelated lines of research during this period.

[35] Heath, 1839: 395.

[36] Ibid.

[37] I saw this trade card on display in an exhibition on Indian science in the Museum of Natural History in South Kensington, London, in April 1982.

First, the earlier metallurgical interest continued. The gradually accumulating basic data in the *Records* and *Memoirs* of the Geological Survey of India and the different district manuals and various government reports prepared primarily in connection with the setting-up of a modern iron industry were incorporated within the broad framework of iron metallurgy in 1864 by John Percy who classified the pre-industrial Indian smelting furnances into three basic types.[38] In this he was followed by another metallurgist, Thomas Turner, who in 1893 published a paper entitled 'The Production of Wrought Iron in Small Blast Furnaces in India'.[39] The main points of this paper were incorporated in his book, *The Metallurgy of Iron*, in 1895.[40] In 1890, there was a concise summary of the available evidence both on the distribution of ores and the pre-industrial smelting practices in the fourth volume of George Watt's *A Dictionary of the Economic Products of India* under the heading 'Iron'.[41] In 1881, Valentine Ball in Part III of *A Manual of the Geology of India* consolidated the evidence primarily from the geological side.[42] In 1880, his *Jungle Life in India or the Journeys and Journals of an Indian Geologist*[43] contained an account of the iron-smelting practices of the tribal group of the Agarias in the Chhotanagpur plateau in eastern India, a field of investigation pursued in detail much later by Verrier Elwin.[44] In 1899 S. A. Bilgami[45] wrote on the iron industry in the territory of the Nizam of Hyderabad and, though Bilgami's account itself is not particularly satisfactory, the paper deserves some attention because Ritter Cecil Von Schwarz who was for some time in charge of the Bengal Iron Company in the 1880s added his comments to it.[46] According to Watt, to him 'belongs the credit of having been the first to show that iron can be successfully smelted in India on European principles'.[47]

The eighteenth- and nineteenth-century interest in pre-industrial

[38] Percy, 1864: 254–70.
[39] Turner, 1893.
[40] Turner, 1908: 324–8.
[41] Watt, 1890: 499–513.
[42] Ball, 1881: 335–414.
[43] Ball, 1880.
[44] Elwin, 1942.
[45] Bilgami, 1899.
[46] Schwarz, 1899: 89–99.
[47] Watt, 1890: 504–5. For a list of Schwarz's other writings on Indian iron, see La Touche, 1917: 464.

Indian iron continued almost unabated till the 1920s. A synoptic account of the east Indian iron industry as a whole with considerable emphasis on the pre-industrial practice was given by Charles Ritter Von Schwarz in 1901.[48] In 1911 Isaac E. Lester's presidential address to the Staffordshire Iron and Steel Institute was on Indian iron.[49] In 1913 Axel Sahlins wrote on the potentialities of India as an iron-producing country and described the Tata Iron and Steel Works.[50] In 1904 F. H. Wynne[51] described the 'native methods of smelting and manufacturing iron in Jabalpur, central provinces'. In 1922–3 Harold Harris summarized the methods of 'native manufacture of wrought iron in small blast furnaces of India'[52] with special reference to Rajdoha in the Chhotanagpur plateau. In 1920 Andrew McWilliam wrote on the manufacture of iron implements and iron at Mirjati in the same area.[53] This paper is considerably significant because McWilliam argued, for the first time, in favour of a continuity of technological tradition between the early and surviving pre-industrial processes in India. The main basis of his argument was a comparison between his analysis of the Mirjati iron and Robert Hadfield's analysis of the compositon of iron of the Delhi pillar of the fourth century A.D. He also pointed out in his paper that 'by careful manipulation especially as to the proportion of charcoal to ore, a steely iron up to 0.9 per cent carbon might be made'. In this context reference can also be made to E. R. Watson's *A Monograph on Iron and Steel Work in the Province in Bengal* in 1907.[54] There is in this monograph a detailed discussion not only on east Indian iron-smelting in the nineteenth century but, most interestingly also, on the blacksmith's craft at the village level—the different articles and the methods of their manufacture. In 1918, N. Belaiew wrote on the nature and compositon of Damascene steel.[55]

[48] Schwarz, 1901.

[49] Lester, 1912.

[50] Sahlins, 1913.

[51] Wynne, 1904.

[52] Harris, 1923.

[53] McWilliam, 1920.

[54] Watson, 1907.

[55] Belaiew, 1918. This is a detailed paper on the Damascene steel and a portion of its 'general conclusions' is worth citing.

1. Under the name of damascene or Damascus steel appeared in Western Europe, during the Middle Ages, the same kind of celebrated Indian steel which was already known in Russia as 'poulad' or 'bulat'.

2. The external characteristic of this steel was its patterned surface-watering, or 'jauhar' (Persian); therefore the Persians called such steel 'paulad jauherder', which means 'steel with

The second feature of this research-phase was an interest in the compositon of early Indian iron objects. Robert Hadfield's paper in 1912 marks a significant attempt in this direction.[56] The immediate subject of this paper was the result of his metallographic examination of three specimens of Sri Lankan iron objects, two from Sigiriya of the fifth century A.D. and the third, an undated but presumably early specimen, from near Kandy. But he also published here the result of his analysis of the Delhi pillar. In his introduction to the paper Hadfield wrote:

> We who live in this modern western world are apt to pride ourselves that we have all the knowledge on this subject of metallurgy, but the facts presented in this paper show this assumption to be incorrect. Whilst information available from the East regarding iron of ancient production is fragmentary, yet undoubtedly a comparatively high state of metallurgical art and knowledge must have prevailed, not only centuries but more than a thousand years ago.[57]

In 1913–14, D. R. Bhandarkar published Hadfield's analysis of an iron wedge discovered in the foundation of the Heliodorus pillar (c. 140 B.C.) at Besnagar, central India.[58] Hadfield found that it contained ·7 per cent carbon and thus could be called steel. This was the first material proof of steel from the early Indian context.

The third feature was a concern with dated iron objects. In 1869 G. G. Pearse published his 'Notes on the Excavation of a Stone Circle

---

a watered surface'. This steel was imported to Russia through Persia and the Caucasus, and became known from the sixteenth century as 'poulad' or 'bulat'; the same steel found its way to Western Europe through Syria and Palestine as damascene or Damascus steel.'

3. The damascene steel was manufactured in India, and apparently this manufacture afterwards spread to some parts of Iran; the origin of the process may be traced back many centuries B.C. According to the author's researches, there were at least three principal methods of producing this steel:

(i) The old Indian, which consisted of producing crucible steel from pure ore and the best kind of charcoal.

(ii) The Persian, when the charge was pure soft iron and graphite, and

(iii) A certain heat treatment, like a very prolonged tempering; this third process is considered by Anossoff as a kind of 'refining process'.

4. During the melting process the greatest care was attached to its intensity and duration; both the Oriental writers and Anossoff pointed out that the best watering could be obtained only on alloys which were kept as long as possible in a molten state and then allowed to cool down gradually with the furnace. When cold, the crucible was opened and the damascene alloy found in the shape of a cake.

5. Such cakes were described by Tavarnier and others, and were brought to this country by Dr Scott. They gave rise to many investigations, notably those of Stodart and Faraday in England, Reaumur and Breant in France, and Anossoff in Russia.

[56] Hadfield, 1912.

[57] Ibid: 134

[58] Bhandarkar, 1913–14.

near Kamptee, Central Province of India'.[59] The exact find-spot was Wurree Gaon. The iron finds were donated to the British Museum. They were found to be of steel and dated about 1500 B.C. or, as it was put, 'the time of Moses'.[60] St John V. Day in 1877 seems to have been the first to draw attention to these finds[61] and since then the Wurree Gaon finds have found mention in various histories of technology.[62] In 1871 Alexander Cunningham furnished positive evidence of the Delhi pillar being of iron.[63] According to some earlier opinions the pillar was of bronze or some kind of mixed metal. Cunningham sent a bit from the rough lower part of the pillar to Dr Murray Thompson of Roorkee Engineering College, U.P., who informed him that the metal was 'pure malleable iron of 7.66 specific gravity'. Cunningham's measurements of the pillar and his estimate of its weight have subsequently been modified but the fact that a solid shaft of wrought iron weighing several tons could be forged and erected as a pillar as early as the fourth century A.D. and could last more or less rust-free since then caught the attention of metallurgists and no praise was high enough for the technical skill which went into its making. James M. Swank, the Secretary and General Manager of the American Iron and Steel Association between 1872 and 1892, considered its forging process a mystery.[64] In 1911 Lester pointed out that this 'until within, say 30 to 40 years ago, was the heaviest piece of wrought iron in the world'.[65] As late as 1930, Henry L. Brose could write, 'only a short while ago it was admitted at a meeting of the Iron and Steel Institute at Birmingham that in spite of modern scientific progress the metal of the 1600-year-old pillar at Delhi . . . was still superior to anything we could produce today; it was freer from intrusions than even Swedish charcoal iron.'[66]

In 1897 Vincent Smith drew attention to another impressive

[59] Pearse, 1869.

[60] Lester, 1912: 8.

[61] Day, 1877: 177.

[62] cf. Forbes, 1972: 248–52.

[63] Cunningham, 1871: 169–70.

[64] Swank, 1892: 68. 'The methods by which the iron pillar of Delhi and the heavy iron beams of Indian temples were forged will probably always remain a mystery. Metallurgists of the present day are certainly unable to understand how these large masses of iron could have been forged by a people who do not appear to have possessed any of the mechanical appliances for their manufacture which are now necessary in the production of similar articles. They could not have been forged by hand.'

[65] Lester, 1912: 6

[66] Brose, 1930.

specimen of early Indian iron, the pillar at Dhar in Malwa, which was found in three pieces, measuring in total about 42 feet long.[67] It was thus longer than the Delhi pillar which is 23 feet 8 inches from top to bottom. Smith dated it in the fifth century A.D. but an early mediaeval date is more likely. In 1912, H. G. Graves wrote a detailed paper on the iron beams of the thirteenth-century Orissan temple of Konarak,[68] a topic further investigated by J. N. Friend and W. E. Thornycroft in 1924.[69] The Wurree Gaon finds, the Delhi and Dhar iron pillars and the temple beams of Konarak became the standard items of discussion in any writing on Indian iron.

The fourth feature was an increasing speculation on the origin and antiquity of iron. Most of the scholars favoured an Indian origin and thus followed the opinion of J. M. Heath of the earlier period. According to Swank, 'India appears to have been acquainted with the manufacture of steel from a very early period'.[70] Lester's comment was: 'India produced the first manganese steel ages before the Western hemisphere was acquainted with the use of iron.'[71] In 1897 H. E. Roscoe and C. Schorlemmer wrote: 'It appears probable that iron was first obtained from its ores in India.'[72] This was also the opinion of John Hawkshaw who, in his presidential address to the British Association for the Advancement of Science in 1875, remarked: 'India well repaid any advantage which she may have derived from the early civilized communities of the West if she were the first to supply them with iron and steel.'[73] In 1886 John Percy commented that 'in India the casting of steel originated'.[74] And as late as 1913, Axel Sahlins could write: 'India is the oldest iron-producing country of which we have definite knowledge.'[75]

There were, however, some exceptions to this kind of opinion. Discussing the knowledge of iron in West Asia and Europe about 1000 B.C., Thomas Turner wrote that 'probably about this time the art of iron-making was carried eastward into India'.[76] Between 1907 and 1910 there was an exchange of opinion between M. Blackenhorn,

67 Smith, 1897.
68 Graves, 1912.
69 Friend and Thornycroft, 1924.
70 Swank, 1890: 8.
71 Lester, 1912: 4.
72 Roscoe and Schorlemmer, 1897: 935.
73 Hawkshaw, 1875: lxxiv.
74 Percy, 1886: 19.
75 Sahlins, 1913: 56.
76 Turner, 1908: 4.

Jules Oppert and Waldemar Belck on the antiquity and origin of Indian iron.[77] According to Blackenhorn, the beginning of iron technology in India could be dated around 1500 B.C. and, according to Oppert, this could be put around 1000 B.C. Belck argued that the earliest iron objects in India could have been brought by the Phoenicians who were familiar with iron earlier than the Indians with whom they had trade contacts. Considering that iron objects were used as the media of barter among the Phoenicians it was, according to him, natural to assume that the Phoenicians used to bring their iron objects to India for barter. He thought it probable that excavations at the old port sites of India would produce Phoenician iron and steel articles. Smith commented in 1912 that the knowledge of iron in India, particularly north India, came from Babylonia. Iron in south India also, according to him, could have come from Babylonia, though in this case the migration was likely to have taken place by sea.[78] In 1926, J. Newton Friend took note of this divergence of opinion. He argued that there could be no objection to a multiple origin of iron and according to him India seemed to be an independent centre of origin.[79]

Albert Neuburger's comments in *The Technical Arts and Sciences of the Ancients* (1930 English translation) are interesting, though somewhat far-fetched. 'We have definite information about the length of time iron was known to the ancient Indians. An iron industry existed in India probably in the year 2500 B.C. and certainly in 1500 B.C. The very fact that the Sanskrit word *ajas* is undoubtedly related to the old Gothic *ais* which later led to the German *eisen*, justifies the assumption that the Indo-Germanic races must have been familiar with iron before they separated.'[80]

It must, however, be emphasized that, despite this continued interest in Indian iron, the archaeologists and historians of India

[77] Belck, 1907, 1908, 1910.

[78] Smith, 1912: 183–4.

[79] Friend, 1926: 150. 'The question has been raised as to whence India derived her knowledge of iron. Babylon is suggested . . ., and it is well established that from very early times iron has been known in what is now Mesopotamia. Others suggest that the Hindoos discovered iron themselves . . . This is very reasonable. There is, of course, no need to attempt to trace back the discovery of iron to one source. Just as the present time discoveries are not infrequently made simultaneously in different parts of the world, so it is reasonable to suppose that in early days, when means of communication were slow and primitive, different nations or tribes may have simultaneously unravelled certain of nature's secrets quite independently.'

[80] Neuburger, 1930: 20–1.

were not by and large interested in the problem. The *Annual Reports* of the Archaeological Survey of India initiated by John Marshall in 1902–3 contain, up to the late thirties when it stopped publication, reports of excavations at different early sites all over India. Hundreds of iron objects must have been excavated from these sites over this period but all that one gets in most of the reports is just a casual reference to 'iron objects'. The rare exceptions are Alexander Rea's reports on the prehistoric antiquities in Tinnevelly in 1902–3 and in Perambair in 1908–9, Bhandarkar's report on Besnagar in 1913–14 and, of course, Marshall's own report on Taxila (containing an analysis of some Taxila specimens by Hadfield) which, however, was not published till 1951. It is also somewhat curious that the numerous iron objects discovered in the south and central Indian graves from the 1920s onwards did not find any particular mention in the early discussions on Indian iron. Even Robert Bruce Foote who made a distinction between the early and later Iron Ages of south India did not have any well-argued date for the antiquity of iron in the south, though once he almost intuitively suggested that iron in India was three thousand years old.[81] Similarly, in his survey of the prehistoric material in *Cambridge History of India*, volume I, in 1922, Marshall did not suggest any specific date for the beginning of iron. However, he took care to emphasize that there was no proof of the arrival of iron in India by sea.

The fifth feature of this phase of research was an emphasis on the literary data. This is evident in a number of writings[82] but, in this regard, it is interesting to look at W. H. Schoff's paper, 'The Eastern Iron Trade of the Roman Empire', written in 1915, in which he shows that the Seric iron of the Classical writers came from India and not from China. 'For the importation into the Roman world from some eastern source of the finest grade of steel then known, there is ample evidence, and it all points toward central India and not China. *Ferrum Indicum* appears in the list of articles subject to duty at Alexandria.'[83] Schoff proceeded to outline the mode and quantity of the transport of steel from India to the Roman world.

> While some of the Indian steel might have been shipped through the Chera ports, it is probably true that most of it went through the port of Barygaza on the Gulf of Cambay, being carried thither by the overland

[81] Foote, 1901: preface.
[82] cf. Ray, 1902.
[83] Schoff, 1915: 230.

> trade route that traversed the great dominions of the Andhra dynasty, 'the inland regions of Ariaca' of the Periplus, thence proceeding westward in native or Arab, and not in Greek or Roman shipping. The product was probably then, as in recent times, bought at the furnaces and the profits of the trade were great enough for the buyers to keep in full themselves without dealing through the third parties. In any case, the total amount shipped westward from India must have been small indeed. In the accounts of the early nineteenth century travellers . . . it is indicated that a single furnace might produce no more than a couple of hundred-weight of steel in a year and 200 tons per year would probably be an outside figure for this export trade.[84]

The Indian literary data in the context of iron were first seriously used by M. N. Banerjee. In 1927 he wrote his first paper on the subject, 'On Metals and Metallurgy in Ancient India'.[85] He went through three types of evidence: primary, which, according to him, included definite mention in the *Rigveda* itself; secondary, which he compiled from the traditions in support of the Rigvedic statements and from post-Vedic literature; and collateral, which was obtained from a comparative review of statements on the subject from diverse authors, both foreign and Indian. According to Banerjee, bronze could not be identified with the qualities possessed by *ayas* in the *Rigveda*. From the Rigvedic references he inferred *ayas* to be the following: 'It was a very hard metal, tough, tenacious, malleable and ductile, one that could be sharpened at the same time into murderous weapons, admit of being easily forged or worked into tools and beaten into desired shapes and sizes.'[86] In 1929, in his paper, 'Iron and Steel in the Rigvedic Age', Banerjee cited the following verse: 'With dried medicinal plants and wings of birds, as also bright stones, the smith awaits wealthy men.' (*RV* 10.72.2) On the basis of this verse which he elucidated with reference to the manufacturing process of pre-industrial Indian steel he described 'how the Rigvedic smith used to manufacture steel in those times':

> He used to manufacture direct from the ore (most probably bright quartzian magnetic stones of iron oxide) in an open hearth (made of mud), adjusted with bellows and fire, and covered over with dried *medicinal plants* such as *Cassia auriculata*; the whole was then heated until the product began to fuse when birds' wings were added for the proper

[84] Schoff, 1915: 236.
[85] M. N. Banerjee, 1927.
[86] Ibid: 128–9.

> and final carburisation of the mass which, at the end of the operation, was frequently hammered into steel.[87]

All that one has to note at this point is that pre-industrial Indian steel was not made directly from the ore. The manufacturing process of wootz, as recorded by Buchanan and Heyne in the context of south India, clearly shows this. In 1932 Banerjee published 'A Note on Iron in the Rigvedic Age'. Here his basic statement was the following: The use of iron and steel in the Rigvedic age can best be proved by showing that the Rigvedic hymns refer directly or indirectly to swords, razor and the quoit ring of iron.'[88]

The sixth and final feature of this phase was an attempt to systematize the whole range of evidence. The most significant attempt was made by Panchanan Neogi in 1914 but mention should also be made of the section on Indian iron in Ludwig Beck's monumental history of iron in 1884.[89] According to Neogi, iron was known to 'the earliest Aryan settlers in India' and he discussed the literary evidence of what he called the Vedic age, the *Brāhmaṇa* age and the Epic age. Archaeologically he mentioned the iron swords, daggers, etc. from Adichanallur in Tinnevelly, which he called prehistoric, iron slag at Bodhgaya (third century B.C.), spearheads and nails at Piprahwa (first or second century A.D.), clamps at Bodhgaya (fourth-sixth centuries A.D.), beams at Bhuvaneshwar, Puri and Konarak (sixth-thirteenth centuries A.D.), pillars at Delhi (fourth century A.D.), guns and cannon (sixteenth-eighteenth centuries A.D.), and the specimens of cast iron from Rangpur (eight-tenth centuries A.D.). The dates suggested by Neogi may not be acceptable in all cases but this was the first attempt to discuss the material evidence of Indian iron through the ages. There was a separate section on what was called 'Matters of chemical interest relating to Indian iron'. This dealt with chemical and metallographic analyses, tensile strength estimate, the varieties of Indian iron deduced from some late Sanskrit texts, the forging and corrosion resistance of early beams and pillars, and the textual data on Indian iron compounds. Finally, there was a discussion on the pre-industrial iron metallurgy, largely based on Buchanan and Heyne.

It may be noted that Neogi was a teacher of chemistry. In this context it is worth recalling that a famous Indian chemist, Sir Prafulla

[87] Banerjee, 1929: 440.
[88] Banerjee, 1932: 364.
[89] Neogi, 1914; Beck, 1884: 203–43, 1008.

Chandra Ray, in his *A History of Hindu Chemistry* showed a concern with the textual material related to Indian iron technology. He cited principally two texts: a treatise dated about 550 A.D. by Varahamihira and a late mediaeval text called *Rasaratnasamuchchaya*.[90] The significance of the citation of such texts is that this citation took place much earlier than any attempt on the part of histcrians to analyse ancient Indian literary data on iron. Moreover, these texts provide some empirical classifications of iron as well as recipes to harden the metal. A recipe mentioned by Varahamihira was the following:

> Make a paste with the juice of the plant *arka (Calotropis gigantea)*, the gelatine from the horn of the sheep, and the dung of the pigeon and the mouse; apply it to the steel after rubbing the latter with sesame oil. Plunge the steel thus treated into fire, and when it is red hot, sprinkle on it water or the milk of horse (camel or goat), or *ghee* (clarified butter), or blood, or fat, or bile. Then sharpen on the lathe.[91]

In his Taxila report published in 1951, John Marshall cited some mediaeval and later references to the excellence of Indian steel.

> Of the fame enjoyed by Indian steel in mediaeval and later times there is no less evidence. Writing in the twelfth century Idrisi says: 'The Hindus excel in the manufacture of iron. They have also workshops wherein are forged the most famous sabres in the world. It is impossible to find anything to surpass the edge that you get from the Indian steel.' In the following century Marco Polo speaks of iron and *ondanique* in the markets of Kerman, and the latter has been recognized by Yule as a corruption of the Persian *hundwany* (Indian steel) which was used for the far-famed sword blades of Kerman. Tavarnier again (1605–80) writes: 'The steel susceptible of being damascened comes from the kingdom of Golconda; it is met with in commerce in lumps about the size of a half-penny cake; they are cut in two in order to see whether they are of good quality, and each makes half the blade of a sabre.'[92]

THE THIRD PHASE (1950–1976)

The first paper to base itself exclusively on the archaeological data was that of D. H. Gordon[93] (1950) who thought that iron could have been introduced in India between 600 and 700 B.C., a premise which possibly made R. E. M. Wheeler[94] (1959) write some time later that

[90] Ray, 1902.
[91] Cited in ibid.
[92] Marshall, 1951; see also Belaiew, 1918, for these references.
[93] Gordon, 1950.
[94] Wheeler, 1959.

iron came to India only with the Achaemenids. Gordon's paper was the first one to discuss the beginning of Indian iron solely on the basis of the archaeological data and as such marked a distinct break with almost everything that had been written on this question since the end of the eighteenth century. It was not entirely a happy break because the earlier writings reflected a research interest much broader in scope than a narrowly archaeological one. The trend initiated by Gordon has continued with little variation up to the present. Gordon's work as well as that of subsequent authors give no indication of the fact that iron was for a long time a significant point of interest in Indian studies, and that it was a field of investigation in which commercial, metallurgical and geological interest combined with that of historians and archaeologists.

The archaeological discovery which strongly militated against the assumptions of Gordon and Wheeler was that of the Painted Grey Ware in the Indo-Gangetic divide and the upper Gangetic valley, first at Hastinapur and then at other sites. The date borders around 1000 B.C. and the discovery of iron in this level was obviously important. Meanwhile, D. D. Kosambi in 1963 depended primarily on the *Suttanipāta* which, he thought, was the earliest of the Buddhist canons. He pointed out clear references to iron ploughshares in this context and, assuming that this obviously wide use of iron in agriculture implied a reasonably early beginning, he put the date of the beginning of this metal in the Gangetic valley, at least in the middle Gangetic valley, at 700–800 B.C. He thought that there was no iron in the early Vedic period.[95]

On the basis of an examination of the archaeological data, B. R. Subrahmanyam's conclusions in 1964 were more detailed: the origin of iron in India was still unknown; the earliest evidence came from the *Doab* of the tenth-eleventh centuries B.C. (his estimate of the date of the Painted Grey Ware); diffusion took place from here to the other parts of India till, around the second century B.C., iron came to be known all over India; the term 'Iron Age' in the sense that the use of iron was widespread could be valid only from the seventh century B.C. He also distinguished three chronological and geographical phases in the early use of iron in India. Between the tenth-eleventh and seventh centuries B.C. iron was limited to the *Doab* and it had a limited practical usage. Between the seventh and fourth centuries B.C., the use of iron spread to eastern U. P., Bihar,

[95] Kosambi, 1963.

Gujarat, and central India. One could also detect in this period an increase both in the quantity and functional types of the iron objects. During and after the fourth century B.C., south India and Orissa could be included among the above areas.[96]

In 1965 N. R. Banerjee published his monograph *The Iron Age in India*. He discussed in detail all the early Indian archaeological contexts in which the iron objects were found and arrived at certain chronological and cultural conclusions. One of these, possibly the most important one, was that the Painted Grey Ware culture of the *Doab*, dated around 1000 B.C., represented the earliest iron-users in India and that this culture could be identified with the Aryans who brought in the iron technology with them from west Asia. Apart from an analysis of the early iron-bearing contexts, there is a discussion on the occurrence of early iron in west Asia and its postulated link with the spread of Indo-European language-speakers. He also discussed the iron objects from different early historic Indian sites and added notes on various miscellaneous items like the early iron technology as revealed at the early historic site of Ujjayini, the method of primitive iron-smelting in India as practised by some modern tribes, the distribution of iron ore in India and finally, what was called 'life in the Iron Age India'.[97]

A study was made by Radomir Pleiner in 1971. The most significant part of his paper is the detailed analysis of the literary data right up to the period of the *Arthaśāstra* with particular attention to the terminology of iron in Sanskrit and Pali literature. His discussion on the Classical references to iron in India is exhaustive unlike his treatment of the archaeological data, which also suffers from the lack of a critical assessment of some of the opinions cited. There is also a brief discussion on the metallographic studies. Pleiner's basic conclusions are: (1) 'iron penetrated into India after the arrival of the Aryans', (2) the iron objects occurring sporadically around 1000 B.C. were without doubt curios, (3) 'the period of penetration of iron into the material culture of India is that of about 800–500 B.C., (4) 'the gradually increasing importance of iron was conditioned since the sixth century B.C. by continuing technical influence coming from western neighbours, from Persia and Media and later from the Greek world', and (5) 'the developed and prospering Iron age' flourished only since 300 B.C.[98]

[96] Subrahmanyam, 1964.
[97] N. R. Banerjee, 1965.
[98] Pleiner, 1971.

Meanwhile, more archaeological data were brought to light after 1965, chiefly in south India, east India, the Northwest and Baluchistan. In each of these areas there were radiocarbon dates for the early iron-bearing levels. These new data along with the old were briefly surveyed by Vibha Tripathi[99] and myself.[100] Accepting the basic diffusionary postulate of Banerjee, Tripathi inferred that by about 1000 B.C. iron had reached the north-western frontier of India and, within the next 200 years, had settled itself in the Gangetic valley. Regarding an early date of about 1000 B.C. for the early iron-bearing level at Hallur in south India, her idea was that it could be explained by an arrival of immigrants from west Asia by sea. My approach was different. First, I delineated the various early iron-using foci in India by the first half of the first millennium B.C. with discussions on their chronology and the broad categories of objects. I then argued that the post-chalcolithic iron-bearing level of central Indian sites like Nagda etc. (and not the Painted Grey Ware culture) should be considered the earliest iron-bearing level in India. Moreover, I pointed out that there was no reason to believe that the dates of the iron-bearing contexts in Baluchistan and the Northwest were earlier than the dates of the relevant contexts in the inner parts of India. As far as the beginning of iron in India was concerned, I thus tried to underline the fact that the available data did not support any *a priori* diffusionist hypothesis. In another paper,[101] I sought to determine the impact of the beginning of iron on the socio-economic situation in India during that period. The data were meagre but I surveyed them area by area and felt that, while the use of iron obviously led to a better agricultural stability, in no case did it bring about a noticeable change in the material prosperity of the people soon after its appearance. No revolutionary role should be attributed to it for the social change preceding the sixth century B.C.

The above-mentioned publications are all basically cultural-historical in their approach without any emphasis on the technological analysis of early Indian iron objects. In fact, there were not many attempts to study early Indian iron objects technologically. In 1963 there were a few metallographic studies of the Delhi pillar[102] and, between 1965 and 1973, there were some analyses of iron objects from

[99] Tripathi, 1973.
[100] Chakrabarti, 1974.
[101] Chakrabarti, 1973.
[102] cf. Bardgett and Stanners, 1963; M. K. Ghosh, 1963; Lahiri, Banerjee and Nijhawan, 1963.

Prakash,[103] Kausambi,[104] Takalghat and Khapa[105] and Mahurjhari.[106] In 1973 H. C. Bharadwaj[107] considered some aspects of early iron technology in India, essentially on the basis of his chemical and metallographical study of a few iron objects and pieces of slag from the early historic level of Rajghat. Not much could be inferred from the limited amount of data except the fact that the objects were of wrought iron. In the same year K. T. M. Hegde[108] carried out an analysis of iron objects from an early historical iron-smelting site near Dhatva in the Tapti valley in Surat district, Gujarat. He aimed at finding answers to a number of purely technological problems: location of the source of the ore, composition of the ore, evidence of ore-preparation, fuel for smelting, composition of the slag, composition of the metal extracted, methods used in the fabrication of objects and the reconstruction of metallurgical techniques. An important part of his analysis was that he pointed out the similarity in technological practices between the early historic iron-smelters and the modern Ghadi Loharias of the same area in Gujarat.

The problem of Indian iron did not generally attract the writers on the socio-economic history of India. As late as 1945, A. N. Bose in his otherwise admirable *Social and Rural Economy of Northern India, c. 600 B.C.–c. 200 A.D.* made only a brief reference to iron, and that too only on the basis of the Buddhist literature, particularly the *Jātakas*.[109] The other general studies on ancient Indian history with their focus on dynastic history did not even mention the problem. There has, however, been some awareness of this problem in recent years. Observations are usually brief, but reference can be made to D. D. Kosambi's notion that a large-scale settlement in the *Doab* was not possible without an effective use of iron, which alone could lead to better clearance of jungle in this area. He also suggested that the growth of the Magadhan empire depended largely on its control over the mineral resources of east India, particularly iron ores.[110] Limitations of such assumptions have been emphasized by Chakrabarti.[111]

[103] Athavle, 1965.
[104] Prakash and Singh, 1968.
[105] Munshi and Sarin, 1970.
[106] Joshi, 1973.
[107] Bharadwaj, 1973.
[108] Hegde, 1973.
[109] Bose, 1945.
[110] Kosambi, 1965: 84.
[111] Chakrabarti, 1973.

THE CURRENT PHASE (1976 ONWARDS)

The current phase possibly began with my definitive study of the problem in 1976. According to this study the present archaeological evidence indicates six early iron-using centres in the subcontinent: Baluchistan, the Northwest, the Indo-Gangetic divide and the upper Gangetic valley, eastern India, Malwa and Berar in central India and the megalithic south India. The archaeological evidence was put in the background of the history of research, the distribution of iron ores suitable for pre-industrial smelting, the data on pre-industrial iron, the literary data and some recent observations on the history of iron outside India. The conclusions arrived at were the following:

> (1) the iron in central and south India is, on the present showing, earlier than the iron in the north-western periphery of India. The central Indian centre seems to be the earliest of the six early Indian centres. (2) Iron seems to have entered the Indian productive system by c. 800 B.C. The literary data alone seem to suggest c. 700 B.C. (3) A look at the list of iron ore areas will show that all these early centres are either in or near the ore areas. The evidence of pre-industrial smelting also comes from almost all these areas. (4) The evidence of pre-industrial smelting and rich ore deposits is very impressive in central and southern India which also seem to show the first evidence of Indian iron. (5) The first Indian iron tool-types do not significantly correspond to the iron tool-types known in west Asia. There is no other demonstrable proof of diffusion during that period from west Asia to the Peninsular block of India. (6) There is an apparent continuity between the early and the contemporary (pre-industrial) traditions of iron metallurgy in India. These points suggest to us that India was a separate and possibly independent centre of the manufacture of early iron. The existing data are admittedly inadequate but the broad indications of this possibility should be clear enough.[112]

Some details of this argument have also been published elsewhere.[113]

In 1978 J. G. Shaffer pointed out the existence of iron artefacts and nodules in the 'Bronze Age' contexts at Mundigak, Deh Morasi Ghundai and Said Qala Tepe in south Afghanistan. He also drew attention to the presence of lollingite nodules in Mohenjodaro and the location of a mature Harappan pot in a deposit at Pirak, which also contained iron artefacts. The data had been brought forward to emphasize a continuum of 'iron awareness' in south Asia. Shaffer

[112] Chakrabarti, 1976.
[113] Chakrabarti, 1977, 1979.

further reviews the Indian Iron Age data and finds my dating of the first iron in Malwa (c. 1100 B.C.) to be on the conservative side. He firmly supports the idea of an indigenous development of south Asian iron technology, an idea which was also endorsed in J. Jacobsen's survey of recent developments in south Asian prehistory and protohistory.[114]

In 1979 M. D. N. Sahi published his analysis of the occurrence of iron objects in Phases Ib and Ic of the chalcolithic Ahar on the basis of the data presented by the Ahar excavation report. Phase Ic yielded 4 arrowheads, 2 chisels, 1 nail, 1 peg and 1 socket and Phase Ib showed 1 arrowhead, 1 ring and 1 slag fragment. There is a C-14 date from Trench X, layer 5, which yields an iron arrowhead—1270 ± 110 B.C. There is no published evidence to suggest that the deposits in which these artefacts occurred were disturbed. On the contrary, their occurrence in 5 trenches (X, C, D, L and E) and in different layers supports their proper stratified locations.[115]

The idea of different early iron-using centres with an indigenously evolving base of iron technology has been accepted with some modifications in two recent interpretations of Indian prehistory and protohistory.[116] B. Allchin and F. R. Allchin now put the first period of iron in the subcontinent as between 1300 and 1000 B.C. but they prefer to correlate the spread of iron in the subcontinent with 'the secondary spread of the Indo-Aryans'. According to them, 'this need in no way conflict with the indigenous population's beginning to exploit local sources of ore to smelt their own ore'.[117] More recently, Gregory L Possehl is attempting to visualize the problem of the beginning of the Indian Iron Age in the overall context of pyrotechnology,[118] but his research results are still more or less unpublished.

Meanwhile, the find of a 'copper' fragment with 66·1 per cent iron and 9·3 per cent copper at Lothal shows that the fragment is really of iron. This proves some Harappan knowledge of iron and should clinch the issue of iron at Ahar.

[114] Shaffer, 1978; Jacobson, 1979.
[115] Sahi, 1979.
[116] Agrawal, 1982; Allchin and Allchin, 1982.
[117] Allchin and Allchin, 1982.
[118] Possehl, 1988.

CHAPTER TWO

Distribution of Iron Ores

The distribution of iron ores in India has been drawing systematic attention since the 1850s. Valentine Ball[1] first summarized the evidence in 1881, and more recently (1954) this has been done by M. S. Krishnan.[2] There are also a few summaries of evidence in individual areas.[3] It is not necessary in this context to refer to the Indian meteorites of which there is, in fact, a large number, but there is no evidence that they were ever used in any way.[4]

Krishnan[5] gives a list of the chief minerals of iron which may be considered as ores. These are magnetite (72.4 per cent iron), haematite (70 per cent iron), turgite (66.3 per cent iron), goethite (59.3 per cent iron), limonite (59.8 per cent iron), siderite (also called 'spathic', 48.3 per cent iron), pyrite (46.6 per cent iron), pyrrhotite (38 + per cent iron), ilmenite (36.8 per cent iron) and greenalite, chamosite, etc. (20–5 per cent iron). Haematite (also called red haematite), brown haematite (consisting of limonite, goethite, etc.) and magnetite have the most significant modern commercial value. Siderite is used only when it occurs in large quantities. Ilmenite is considered more as an ore of titanium than of iron. Pyrite can be used as an ore if its sulpher content is first eliminated by oxidation through roasting.

According to Krishnan,[6] the iron ores of India are classified

[1] Ball, 1881: 335–414.

[2] Krishnan, 1954.

[3] cf. Dunn, 1942; Krishnan, 1951: 138–53; Krishnan and Aiyengar, 1954; Roy Chowdhury, 1955: 37–41; Roy, 1959: 195–201; Hunday and Banerjee, 1967: 175–90.

[4] cf. Krishnan, 1954.

[5] Krishnan, 1954: 11–13.

[6] Krishnan, 1954: 102–3.

into three major groups according to their origin. The banded ferruginous formations of the pre-Cambrian age constitute the most important group. The majority of its larger deposits is composed of the unmetamorphosed type in which the ore-bodies are derived from the concentration of iron in the original haematite-jasper formation. In the metamorphosed type, these ores are converted into banded quartz-magnetite rocks in which the magnetite is derived from the original haematite. The sedimentary iron ores of sideritic or limonitic composition form the second group which may be observed in the Iron-stone Shale deposits of the Lower Gondwana age of the coal-fields of Bihar and West Bengal and in the ferruginous beds of the Tertiary formation of some areas of the Himalayas and Assam. The lateritic ore, the third major group, is the result of sub-aerial alteration of the iron-bearing rocks like gneisses, schists, basic lavas, etc. under humid tropical conditions. The lateritic spread is common in India and covers large stretches of the Deccan traps, the gneisses in the Western Ghats, the schistoise rock formations of many areas, the impure lime-stone formations, etc. The iron content of this group is low, only 25–35 per cent. Among the other groups may be mentioned the apatite-magnetite rocks of the Singhbhum copper belt, the titaniferous and vanadiferous magnetites of south-east Singhbhum and Mayurbhanj, and the fault and fissure fillings of haematite, a good example of which may be observed in the Kurnool area of Andhra Pradesh.

While reviewing the distribution of iron ores of various geographic areas of India as a backdrop to her pre-industrial smelting through the ages it is necessary to emphasize a basic point at the outset. This point was made by T. H. Holland and L. L. Fermor as early as 1910 but has not always been remembered since then.

> Iron smelting was at one time a wide-spread industry in India, and there is hardly a district away from the great alluvial tracts of the Indus, Ganges and Brahmaputra, in which slag-heaps are not found. But the primitive iron-smelter finds no difficulty in obtaining sufficient supplies of ore from deposits that no European iron-maker would regard as worth his serious consideration. Sometimes he will break up small friable bits of quartz-iron-ore schist, concentrating the ore by winnowing the crushed materials in the wind or by washing in a stream. Sometimes he is content with ferruginous laterites, or even with the small granules formed by the concentration of the rusty cement in ancient sandstones.[7]

[7] Holland and Fermor, 1910: 99.

Regarding the occurrences of quartz-iron-ore schist in India Holland and Fermor's comment is that they are so common in India that 'newly recorded instances are generally passed over as matters of very little economic interest'. The point is that a survey of the distribution of Indian ores on the basis of the Geological Survey of India reports, however exhaustive it may be, may not be wholly representative of the sources open to a pre-industrial iron-smelter.

The distribution begins right from the north-western and northern limits of the subcontinent. In Sind[8] the most important source is the passage beds between the Kirthar and Ranikot groups, north-west of Kotri, especially near Laniyan and east of Bandh Vera. The beds, sometimes 15–20 feet thick, are not often adequately ferruginous, though there are significant occurrences of magnetite, and red and brown haematite in places. The beds continue west and south-west of Jhirak, but with a lesser iron content. There are ferruginous rocks also at the base of the Manchar group where the latter rests upon the Kirthar limestone near Bandh Vera and at the base of the Laki range. Moving towards the Northwestern Frontier province, one finds iron ores in the form of an earthy haematite to the south-west of Bannu. Ball reported that this was locally in great demand for making nails, cooking utensils, etc.[9] In the Peshawar area a black magnetic iron sand was collected from the neighbouring Bajaur and used for the markets of both Peshawar and Kabul. One of the purposes for which it was smelted in Peshawar was for the making of gun barrels.[10] Magnetite and haematite occur in the Hazara country and there is a clay iron stone in the Tertiary rocks of the Bolan.[11] A limonitic ore exists in the hills of Waziristan and there are also records of its pre-industrial smelting.[12] One particular area called Kanigaram is specifically noted because of its local reputation as an iron-smelting centre, visible in the form of furnaces, slag and stores of ore in the villages. In the extreme north, iron ores of different types—fairly rich segregations of magnetite, and haematite, limonite and other sedimentary rocks—are widely distributed in different parts of Kashmir which also had pre-industrial smelting furnaces in operation

[8] Ball, 1881: 402.
[9] Ibid: 404.
[10] Ibid: 404.
[11] Ibid: 402.
[12] Ibid: 403.

in the nineteenth century. The main iron ore areas are Gangani, Ladda, Matah and Khandli.[13]

An earthy haematite is found both in the Salt range and the Kot Kerana hills of Panjab,[14] but more significantly in the latter area. An attempt was made here in the nineteenth century to produce iron on modern commercial lines. In the Panjab foothills there is an abundant supply of magnetic and micaceous iron ores in Kangra, which was used by the local pre-industrial iron-smelters. Similar ores also occur at Shele to the east of Simla.[15] The deposits of Mandi[16] consisting of banded magnetic schists and micaceous haematite schists are considerable—in fact, 60 million tons according to a recent estimate. Not much is known about the iron ores of Kulu,[17] but a report exists on its pre-industrial smelting which included the preparation of saucepans, large boilers for sugar, and even swords. Magnetite, probably mixed with specular ore is found in Sirmur which had a modern blast furnace in operation as early as the 1880s.[18]

Continuing along the same Himalayan belt to the east, there is a significant occurrence of iron ores in Nainital, Almorah and the Garhwal areas of modern U. P., or broadly, the area known earlier as Kumaun. Red and brown haematites are the principal ore-types. There was a local pre-industrial iron industry and once there was a considerable interest in the modern use of the ores of Kumaun. Ball sardonically remarks that 'the amount of iron which has been manufactured bears but a small proportion to the number of reports which have been written upon the subject'.[19]

In Panjab outside the Himalayan zone the important deposit of iron ore is in Patiala, once a centre of pre-industrial smelters.[20] The ore-types are basically brown haematite and magnetite. Iron is pretty generally distributed in the state of Rajasthan.[21] There are noteworthy deposits in Alwar, Jaipur, Udaipur and Ajmir, and there are reports of ancient workings also from Bharatpur, Bundi, Jodhpur and Kota. Micaceous haematite-schists and a mixture of magnetite and ilmenite seem to characterize the deposits in Alwar. In 1873 there was a

13 Ibid: 404.
14 Ibid: 404–5.
15 Ibid: 405.
16 Krishnan, 1954: 157–8.
17 Ball, 1881: 405.
18 Ibid: 407–8.
19 Ibid: 406; Krishnan, 1954: 98–100.
20 Ball, 1881: 406; Krishnan, 1954: 177.
21 Roy, 1959: 195–201; Krishnan, 1954: 177–9; Ball, 1881: 395–7.

report of about 30 pre-industrial furnaces working in this area. The traces of old mines are supposed to be very much evident in Alwar, particularly at Bhangarh and Rajgarh. The open-cast mines of Bhangarh on the summit of a hill are said to be several hundred yards long and 20–30 yards wide, and connected with short irregular adits. Limonite and haematite are the principal types of ore in Bundi as they are in Udaipur. There are traces of early working and mining in both these areas. In Jaipur haematite seems to occur in plenty in various formations, along with the irregular masses of highly ferruginous rock. In Jodhpur the ore is largely a mixture of haematite and magnetite associated with quartz schists. The deposits in the rest of Rajasthan also seem to conform to these main types. The traces of early working and mining are eqully abundant, though they have not been systematically surveyed and studied.

Iron exists also in the adjacent state of Gujarat.[22] In the Balsar and Pardi areas of Surat iron ores are reported to occur, and there are reports of early slag-heaps in these and other parts of the district. Also, magnetic sand is said to accummulate at the mouths of the rivers, particularly at the Dumas side of the Tapti. Iron ores of considerable richness are said to be found in Panchmahal, and also in Kaira, Rewa Kantha and Ahmedabad, particularly if the pre-modern slag-heaps in all these areas are any indication. In Kathiawar the ores were in some cases obtained by sinking pits in the alluvium and thus reaching the buried lateritic rock. Pre-industrial furnaces were in operation in Kathiawar in the nineteenth century. In Kutch also the ore seems to be lateritic and there was a pre-modern smelting tradition in Kutch.

The whole of central India is iron country *par excellence*.[23] In the former Madhya Bharat which broadly conforms to the tract of Malwa in the modern state of Madhya Pradesh, iron occurs in the geological formations of laterite, the Vindhyan system, the Gwalior series and the Bijawar series. There are extensive cappings of laterite rich in iron in Bhilsa, Ujjain, Shajapur, Shivpuri and Mandasore. The iron in the Vindhyan system is found principally in Mandasore and that in the Gwalior series in Gwalior. The iron in the Bijawar series primarily covers Indore, Jhabua and Dhar. The basic ore-type seems to be haematite which is also found in Lalitpur, Hoshangabad, Narsinghpur and Nimar. The area around Gwalior is said to possess several remarkably rich deposits, particularly at Par, Mangor, Maesoora

[22] Ibid: 399–402.

[23] Roy Chowdhuri, 1959: 37–41; Ball, 1881: 383–93; Krishnan, 1954: 139–57.

and Santao which all possess extensive evidence of pre-industrial mining. The mines have been reported to be generally deep shafts from which smaller galleries were extended. At Salda in Lalitpur a pure haematite is found, and a soft iron smelted from it was once exported to the neighbouring areas. The ores of Omarpani in Narsinghpur are said to have been quite famous once. The mines consisted of pits and burrows which followed the courses of the ore to a depth of 40–50 feet. The ores collected at Omarpani were taken to Tendukhera, about 2 miles away, for smelting. In Nimar there are old mines at Barwai where the ore is a pure red haematite. In fact, there is no dearth of evidence of pre-industrial smelting in the whole of Malwa territory.

In the rest of central India or the modern Madhya Pradesh the principal iron-bearing deposits are in Bastar, Chanda, Drug and Jubbulpur. Bilaspur which is slightly to the east also possesses iron and the same may be said about the other areas like Banda, Mandla, Bundelkhand and Rewa. The Bailadila, Rowghat and Kanker deposits in Bastar are among the important sources of raw material of the modern Indian iron and steel industry, and their magnitude may be inferred from the fact that the Bailadila deposit alone is estimated to possess 3,600 million tons of high grade ore in addition to large quantities of lower grade ore. Haematite of varying composition and thus of varying richness of iron content seems to be the main ore-type in these deposits. The ore-type is haematite in Durg, but with small quantities of magnetite in places. There are also occurrences of hydrated oxides and lateritic products at or near the surface of the outcrops. The ores mainly occur on the Dhalli and Rajhara ridges which rise to a general height of 400 feet from the level of the plain and continue for about 20 miles. The ores, again mainly haematite, occur on the northern part of Chanda where they rise in the form of hillocks. The principal localities are Lohara, Pipalgaon, Asola and Dewalgaon. A hill named Khandeshwar near Dewalgaon is about 250 feet high and its entire mass is laden with ore. At one point it was thought that this hill alone might furnish the whole of India with iron. There is an extensive body of literature on the iron ores, traces of pre-industrial smelting, and the possibilities of modern manufacture in Chanda.

The estimate of the total iron deposit in Jubbulpur is 100 million tons and, according to another estimate, iron is found at more than a hundred places in the district. The principal occurrence is in its north-western part. The ore is mainly haematite, derived from many sources; at one place it is known to occur in the form of black sand. The most extensively worked cluster of old mines was noted at a place called Majgaon, and at Palle in the same area, the excavations for ore were (in 1872) 100 yards long, 30 yards wide and 50 feet deep. A pre-modern iron industry existed in the neighbouring areas of Rewa, Bundelkhand, Banda, Mandla and Bilaspur which all had their own sources of iron. In Maharashtra[24] note should be taken first of Konkan where laterite is widely distributed and was used by the pre-industrial smelters in different pockets like Ratnagiri, Kolhapur and Mahabaleshwar. The mines, particularly in the Kolhapur area, are said to have been shallow pits, never more than 8–10 feet deep. Considerable deposits, however, occur in Goa in the Dharwar formations.

Southwards in Mysore,[25] iron ores are fairly extensive and geologically belong mainly to the sedimentary group, associated with banded haematite-quartzites of the Dharwarian age, and partly to the deposits of magmatic origin which are titaniferous ores. Apart from these the lateritic spread of south Konkan should be considered a potential source. The early iron workings of Mysore have been made justly famous by the reports of Buchanan and Heyne. In the Tumkur district iron ore is said to be abundant in the Chikayakanhalli hill and magnetic sand is brought down by the streams from the rocks at Madgiri and Kostagiri. In the area of Mysore itself, iron ore is abundant in the rocky hills and the same is true of Shimoga, Kadur and Chitaldurg. Iron ores in the form of haematite form ranges of hills in Chitaldurg and in Kandur they are largely obtained and smelted along the hills east of Baba Budan and those around Ubrani. While writing on the annual quantity and value of iron smelted in the former princely state of Mysore in 1897, B. Lewis Rice[26] pointed out that Shimoga district produced the greatest amount, followed by Chitaldurg, Kolar and Kadur districts. In the other districts, the following areas were the principal centres of the industry: Magadi, Chikayakanhalli, Gubbi, Heggadadevankote and Malvalli.

In the north of Kerala (formerly called the Malabar districts) iron ores are abundant and comprise mostly magnetite and laterite. Magnetite occurs in the form of bands in the metamorphic rocks and of black sands derived from the bands. The amount of pre-industrial

[24] Ball, 1881: 398–9; Krishnan, 1954: 137–9.
[25] Ball, 1881: 352–6; Krishnan, 1954: 173–6.
[26] Rice, 1897: 530–5.

operations was quite considerable; in 1854 more than 100 furnaces were in operation. In the south of Kerala, or the former Travancore, the ore is chiefly derived from the lateritic spread, though black magnetic sand is frequently cast upon the beach in enormous quantities along the southern section of the coast. Shenkotta was the principal centre of pre-industrial smelting in this area.[27]

In the adjacent state of Madras, or modern Tamilnadu,[28] the ores occur as far south as Tinnevelly where they are chiefly magnetite and laterite. The ore used in the local pre-industrial furnaces was magnetite in the form of magnetic iron sand. In Madura the ores were derived from the lateritic conglomerates, though ore deposits of magnetite were also observed. The pre-modern industry of this area was found extant even in the nineteenth century and large heaps of slag testified to its former existence. The deposits in parts of Salem and Trichinopoly are geologically significant enough to have drawn a separate monograph on this area by Krishnan.[29] The ore-type is basically magnetite, with a much less amount of haematite. Mounds of slag were once noted in Trichinopoly and the smelting operations were current, though not on an extensive scale, in Salem in the nineteenth century. According to Ball, 'the development of magnetic ores in the Salem district is among the most remarkable facts connected with the geology of India, whether the extent, thickness, or number of the beds be considered'.[30] In Coimbatur where the smelting furnaces were mentioned by Buchanan, magnetite of good quality is said to occur. The ores, mainly haematite and specular, occur also in the Nilgiris where the most important mass is near Kotagiri. The other areas of occurrence in Madras are Chingleput, Pudukottai and north and south Arcot, each of which has had a tradition of pre-industrial smelting. Magnetite is the principal ore-type of these areas.

The spread of the iron ores is no less extensive in the modern state of Andhra Pradesh.[31] The regions which are important are Cuddapah-Kurnool, Guntur, Bellary, Nellore, the districts of West Godavary and Krishna, Vizagapatam and Hyderabad. In Cuddapah the significant ore-bearing deposits are at Chabali, Pagadalapalle, Pendlimarri and Mantapampalle. The basic ore type seems to be haematite derived from the ferruginous quartzite formations locally

[27] Ball, 1881: 346, 350.

[28] Ibid: 346–50; Krishnan, 1951: 138–53; Krishnan and Aiyengar, 1954.

[29] Ibid.

[30] Ball, 1881: 348.

[31] Ibid: 357–62; Krishnan, 1954: 158–61, 162–5, 166–8.

enriched to iron. At Pendlimarri the quartz veins contain specular ore also. Pre-industrial smelting seems to have been once generally distributed all over Cuddapah. There is a reference to a series of iron-smelting villages lying along the eastern side of the Khundair valley. 'In other parts of the district there were also furnaces, and itinerant blacksmiths, carrying with them the implements of their trade, wandered over the district seeking for employment.'[32] The nature of the ore and its deposition do not vary in Kurnool. The best ore is said to be found in the Gunnygull range near Kurnool town, which contains veins of pure specular ore. There is also no lack of the occurrence of ferruginous quartzites in the district. There were a number of smelting villages in Kurnool. Limonite occurs in Guntur which also possesses some bands of quartz-magnetite rocks and a tradition of pre-industrial smelting. Ferruginous quartzites and ores are noted in the Sandur range of Bellary. Several other sources exist in this district and, in some cases, the ore is said to be manganiferous. Reports exist on pre-industrial smelting in Nellore which possesses both magnetite and haematite-bearing schists. There was pre-industrial smelting also in the districts of West Godavari and Krishna, for which the ore was principally derived from the limonite patches in the Rajamahendri and Golapilli sandstones. Haematite and magnetite both occur in the area of Vizagapatam, again with a tradition of pre-industrial smelting. At a place called Chitra in Vizagapatam, Ball[33] found traces of extensive mining: a deep trench was found along the outcrop for nearly a mile. Iron ores consisting of a mixture of magnetite, haematite and limonite are extensively found in the Dharwar formations of Hyderabad, besides which the laterite cappings of the Deccan trap provide some of the poor grade ores. The most famous ore of Hyderabad is, of course, magnetite used for manufacturing the wootz of the Damascus blades.

To the east, in Orissa there are enough lateritic cappings outside the Mahanadi-Baitarani delta, and for the early smelters of Orissa these deposits alone could have been an ample source of iron.[34] The deposits which need particular attention are in Bonai-Keonjhar, Talcher, Sambalpur and Mayurbhanj. In Bonai-Keonjhar the ore-bodies are haematite, and they are a continuation of the deposits of Singhbhum in Bihar. According to Krishnan 'the most important

[32] Ball, 1881: 357.
[33] Ibid: 360.
[34] Ibid: 361–2; Krishnan, 1954: 121–37.

group of deposits of iron ore in India occur in south Singhbhum and the adjoining districts of Bonai and Keonjhar within the area bounded by latitudes 21°40′N and 22°20′ N and longitudes 85°5′ E and 85°32′ E.'[35] The estimate of the total deposit of high grade ore in this region is 8,000 million tons. The deposits in Mayurbhanj, particularly those of titaniferous magnetite, are also a continuation of the deposits in Singhbhum. Apart from this, Mayurbhanj possesses large haematite deposits. The pre-industrial smelting operations of Sambalpur and Talcher drew considerable attention in the nineteenth century, though the deposits are less in this area compared to the Bonai-Keonjhar-Mayurbhanj belt.

A. Hunday and S. Banerjee have summed up the mode of occurrence and the areas of iron deposits in the state of West Bengal.

> Occurrences of iron-ores are reported both from the western and northern districts of the state. Haematite, magnetite and titaniferous iron-ores are found sporadically in the Archaean tracts of Purulia, Midnapur, Bankura, Birbhum and Darjeeling districts; sideritic and limonitic ores in the Iron-stone shales of the Raniganj coal-field; haematitic beds within Siwalik sandstones of the Darjeeling district; lateritic iron-ores in the Purulia, Midnapur, Bankura, Burdwan and Birbhum districts; and magnetic iron-bearing sand in some river beds in the Purulia district.'[36]

The list, in fact, covers all the major areas outside the Gangetic alluvium. One of the most important centres of pre-industrial iron-smelting in West Bengal was Birbhum. There is an extensive body of literature on the pre-industrial iron-smelting of Birbhum and its adjacent areas.

Modern Assam is not exactly known as an iron-producing state but that it was once so is amply confirmed by the earlier *Geological Survey of India* and other reports.[37] The principal ores are clay-stones from the coal-measures which occur in nodules of various sizes and also in thin beds interstratified with shales and sandstones, an impure limonite from the sub-Himalayan strata. According to a report, in the middle of the nineteenth century there were 3,000 smiths and smelters in upper Assam. Another important area was the Khasi-Jaintia hill. The principal locality in which the mines were worked is near Chera Punji and this is said to extend for a distance of 6 miles

[35] Ibid: 127.
[36] Hunday and Banerjee, 1967: 175.
[37] Ball, 1881: 412–4; Krishnan, 1954: 109–11.

from east to west and of 2 miles from north to south. The ore was basically a titaniferous, magnetic iron sand.

Bihar[38] possesses her major iron resources in the Chhotanagpur plateau, largely in the areas of Hazaribagh, Singhbhum, Ranchi and Palamau. These areas are important even now and a bare reference to their ore-types should suffice. The main ore-type seems to be haematite but there are magnetite, limonite and titaniferous ores as well. According to J. A. Dunn,[39] 'indigenous iron-smelters were perhaps more flourishing in Bihar than in any other part of the country.' Iron occurs also elsewhere in Bihar, for example, in Mungher and the Santal Parganas, both in the laterite and in thin bands in the schists of the hills.

Apart from the deposits in Kumaun-Garhwal the source of iron in modern U.P. seems to be in the Mirzapur district where the principal ores are magnetites found in bands in the sandstones of the Barakar age. Magnetic sand also is supposed to be common in the streams of this area.[40]

Two points should clearly emerge out of the foregoing brief survey: (1) iron suitable for pre-industrial smelting, if not always for modern industries, occurs in almost all the areas of the subcontinent outside her major alluvial stretches, and (2) wherever iron occurs, there is almost invariably some trace or tradition of pre-industrial smelting. It would have been quite unnecessary to stress these points if they had been noticed by some of the modern writers on early Indian iron. For instance, N. R. Banerjee writes that 'the deposits near Narnaul in Patiala and in Mandi mark the northernmost major deposits known in India' and that 'it would have been scarcely possible for the Aryan settlers of India to start the manufacture of iron objects, until not only had they reached the area but discovered the rich deposits of the ores of Narnaul'.[41] Considering that there was a flourishing pre-industrial iron-smelting tradition with enough local ores to maintain it in the north-west as far as Peshawar and the Waziri hills, such statements are pointless and misleading. R. J. Forbes in his distribution map of 'Some of the Most Important Iron Deposits of the Ancient Near East'[42] refers only to Hyderabad in India whereas

[38] Krishnan, 1954: 116–37.

[39] Dunn, 1942: 142.

[40] Ball, 1881: 393.

[41] N. R. Banerjee, 1965: 190.

[42] Forbes, 1972: figure 30.

he could have referred to the entire Peninsular block of India. Radomir Pleiner realizes[43] that 'it is not necessary to confine ourselves only to the locations of the most important raw material resources in the country when attempting to trace Indian metallurgy to its point of origin. Particularly in the case of iron, the doubtlessly small-scale activities need not have been bound to the excellent ore-fields alone.' However, in his map of the occurrence of iron ore deposits he chooses to show only the major source areas. On the basis of the data discussed above we provide the following list.

A List of Iron-bearing Areas and Localities in India

1. *Sind*: Kohistan—Ranikot area (magnetite, red and brown haematite).
2. *Baluchistan*: Bolan (clay iron-stone).
3. *Northwest Frontier zone*: Bajaur (black magnetic iron sand), Bannu (earthy haematite), Hazara (magnetite and haematite) and Waziri hills (limonite). Pre-industrial smelting.
4. *Kashmir*: different parts of the state (magnetite, haematite, limonite and other sedimentary ores). Pre-industrial smelting.
5. *Salt Range and Kot Kerana hills*: earthy haematite.
6. *Panjab and U. P. Himalayas*: Kangra (magnetic and micaceous), Shele near Simla (magnetic and micaceous), Mandi (magnetite and haematite, micaceous schists), Sirmur (magnetite probably mixed with specular ore), Kulu (?), Nainital, Almorah, Garhwal (red and brown haematite). Pre-industrial smelting.
7. *Patiala*: brown haematite and magnetite. Pre-industrial smelting.
8. *Rajasthan*: Alwar, Jaipur, Udaipur, Ajmir, Bharatpur, Bundi, Jodhpur, Kota (basically haematite, magnetite, limonite). Pre-industrial smelting. Also traces of mining.
9. *Gujarat*: Surat, Panch Mahal, Rewa Kantha, Ahmedabad, Kutch, Kathiawar (lateritic ore in Kutch-Kathiawar; ore unspecified in the rest). Pre-industrial smelting. Also numerous slag-heaps.
10. *Central India*: almost the whole of it; lateritic ore in Bhilsa, Ujjain, Shajapur, Shivpuri, Mandasore; the ore in the Vindhyan system principally in Mandasore and that in the Gwalior series in Gwalior. The ore in the Bijawar series in Indore, Jhabua and Dhar. Also, haematite in Lalitpur, Hoshangabad, Narsinghpur, Nimar. Important deposits in Bastar, Chanda, Durg. Other areas: Jubbulpur, Bilaspur,

[34] Pleiner, 1971: figure 1.

Banda, Mandla, Rewa, Bundelkhand. Pre-industrial smelting. Also, traces of mining.

11. *Maharashtra*: Konkan laterite; also, Ratnagiri, Kolhapur, Mahabaleshwar. Important deposits in Goa in the same area. Pre-industrial smelting.

12. *Mysore*: laterite of south Konkan, haematite-quartzites and titaniferous ores; also magnetic sands. Areas: Tumkur, Mysore, Chitaldurg, Kadur, Kolar, Shimoga. Pre-industrial smelting.

13. *Kerala*: magnetite and laterite, also black magnetic sand. Malabar and Travancore. Pre-industrial smelting.

14. *Madras*: magnetite, haematite, laterite; also specular ore. Areas: Tinnevelly, Madura, Salem, Trichinopoly, Coimbatur, Nilgiris, Pudukottai, Chingleput, north and south Arcot. Pre-industrial smelting. Slag-heaps.

15. *Andhra*: magnetite, haematite, limonite, etc. Areas: Cuddapah-Kurnool, Guntur, Bellary, Nellore, West Godavari, Krishna, Vizagapatam, Hyderabad. Pre-industrial smelting. Traces of mining.

16. *Orissa*: laterite outside the Mahanadi-Baitarani delta, magnetite, haematite of Bonai-Keonjhar, Talcher, Sambalpur, Mayurbhanj. Pre-industrial smelting.

17. *West Bengal*: haematite, magnetite and titaniferous ores in the Archaean tracts of Purulia, Midnapur, Bankura, Birbhum, Darjeeling; siderite and limonite in Raniganj coal-field; haematitic quartzite in Jalpaiguri; haematitic beds in Darjeeling; laterite in Purulia, Midnapur, Bankura, Burdwan, Birbhum; magnetic sand in Purulia. Pre-industrial smelting.

18. *Assam*: clay iron-stone and an impure limonite in upper Assam; titaniferous magnetic iron sand in the Khasi-Jaintia hill. Pre-industrial smelting.

19. *Bihar*: different ore-types in the entire Chhotanagpur block; also, Mungher. Pre-industrial smelting.

20. U. P. (except Nainital-Almorah-Garhwal): magnetite and magnetic sand in Mirzapur.

CHAPTER THREE

The Nature of the Archaeological Evidence

The purpose of this chapter is to put forward an idea of the regional data on iron on the basis of the representative excavated sites and the technical analyses of at least some of the excavated artefacts. This is not intended to be an exhaustive survey but the basic character of the evidence will be set down regionwise, hopefully with some clarity. In 1976 I argued that there were six early iron-bearing foci in India—Baluchistan, the Northwest, the Indo-Gangetic divide and the upper Gangetic valley, eastern India, Malwa and Berar in central India and the megalithic south India.[1] Moreover, in view of M. D. N. Sahi's analysis of the evidence of iron from Ahar in south-east Rajasthan,[2] that area also should now be added to the list. Emphasis will be laid primarily on the above-mentioned regions but reference will be made to the data from the other geographical contexts as well.

I. REGIONAL SURVEYS

1. Baluchistan

Two types of sites are associated with the beginning of iron in Baluchistan: the type of site represented by the excavated settlement of Pirak and the 'cairn-burials'. There is no evidence that these two types of sites were related in any way, chronologically or otherwise.

Pirak

The site of Pirak is in the Kachhi plain of Baluchistan. There is no

[1] Chakrabarti, 1976.
[2] Sahi, 1979.

major break in its sequence and the number of its occupational levels (numbered from the top) is eleven. Iron appears in the level 6 but in an obviously limited quantity. Its use increases in the levels 4 and 3. Several two-winged arrowheads have been found, one in association with a blacksmith's furnace. Serrated stone blades from the earlier chalcolithic levels continue in use. Two of the earlier three chalcolithic pottery types which are in the bichrome category continue to occur, though the bichrome painting style is said to be cruder. A new ceramic element is a grey or black pottery, wheelmade and often burnished. Its common shapes are carinated bowls and dishes, occasionally with horizontal stud-handles.

About 1,000 square metres of the level 4 are excavated. The basic component is a set of rooms within an enclosing wall. The niches and the interconnecting doors had wooden lintels. There were also ovens and fireplaces and a few storage jars half-buried in the ground. In the third level the houses were rebuilt and the greater number of fireplaces, ovens and artefacts may denote an increased level of craftsmen's activities. There are some terracotta seals with compartmented designs and beads decorated with zigzags and circles. An important feature is a large-scale preparation of bone points, mostly of antler and often decorated with an incised circlet on each side.

The upper levels (the levels 1 and 2) are mostly a mass of ash, crude pots and ovens, but one notes a well which is lined with wedge-shaped burnt bricks and also a 5cm high, well-finished terracotta head.[3] There are about 14 radiocarbon dates from various levels of Pirak[4] and, for the iron-bearing horizons, the calibrated date seems to hover around 800 + B.C.

Two points may be noted about the beginning of iron at Pirak. First, this does not bring about any break in the cultural sequence. There is a significant element of continuity in the use of the earlier pottery types and stone blades. Secondly, the number of the so-called fireplaces and ovens suggests smelting activities.

The Problem of the Cairn-burials

The only reason why this problem has to be examined in some detail is because these cairn-burials have been related by a number of scholars to the beginning of iron in India. The cairn-burials of Pakistani

[3] Jarrige and Enault, 1973.
[4] Possehl, 1988.

Baluchistan and Makran, which are distributed also in Persian Baluchistan and Makran and extend in that country up to Fars and Kirman, were first noted by Major Mockler in 1876 in the lower Dasht valley near Gwadar on the Makran coast.[5] Mockler's four sites—Damba Koh, Darmani Ban, Jiwanri and Gatti—were re-explored by Aurel Stein[6] (1931) who also discovered similar burials in the Zhob-Loralai area of north-east Baluchistan (1929).[7] In two later expeditions (1936, 1937),[8] Stein traced their distribution up to the Kirman and Fars areas of Iran. Apart from limited work at Sar-i-Asiab near Kirman by C. C. Lamberg-Karlovsky and James Humphries in 1968, there has been no basic fieldwork on these burials since the work of Stein.[9]

The total number of sites is about 45, about 20 of them being in Pakistani Baluchistan and Makran.[10] The total number of cairns must be several thousand; Damba Koh alone is estimated to possess 2,000 of them.[11] The concentration of individual burials seems to be more in the Dasht plain on the border of Iranian and Pakistani Makran than elsewhere. The Dasht sites are coastal plain sites; one of them, Take Dap, is right on the sea.[12] The quantity of fish-bones in one of the Take Dap cairns and the kitchen-middens nearby is suggestive of the livelihood of the cairn-burial people of this area. Though Bushire on the Persian Gulf and Tiz on the Chahbar Bay are sometimes called cairn-burial sites,[13] there is no positive proof of this. The general setting of the sites is in most cases a hill slope or crest, never far from a source of water, and in some cases there are indications of stone and mud-built dwelling structures nearby.[14] There is little doubt that the stone slabs for the cairns were collected locally. It is worth noting that in Pakistani Baluchistan, the sites are primarily in the lower reaches of the Dasht and the Kej valley in the south and the Zhob-Loralai area in the north-east. The sites in the intervening areas of Kharan, Jhalawan and Sarawan seem to be comparatively few. No

[5] Mockler, 1876.
[6] Stein, 1931.
[7] Stein, 1929.
[8] Stein, 1936, 1937.
[9] Lamberg-Karlovsky and Humphries, 1968.
[10] See Chakrabarti, 1979.
[11] Ibid.
[12] Ibid.
[13] Ibid.
[14] Ibid.

site seems to exist in the valley of Quetta or in Persian or Afghanistan Seistan. This may reflect a basic distribution pattern or merely a lack of exploration.

The cairns seem to have belonged to four types. The first is what Mockler called 'square' and Stein, 'sub-rectangular'. The size varied; at Damba Koh the average one in Stein's account measured 4–5 ft. high and 5–6 ft. square. The interior was filled with loose earth containing the burial deposit, and was roofed with stone slabs. Mockler specifically mentioned doors with stone lintels. Stein only referred to a small opening on the side, which marked the entrance. Mockler also mentioned that in some cairns of this type, the side walls narrowed towards the top to form part of the roof. Stein did not mention this. At Hajjiabad in the Kirman area, Lamberg-Karlovsky and Humphries noted 'walls stepped inwards in a crude imitation of a corbelled vault'.

The second type was simply a circular pile of stones with a patch of earth in the centre, which contained the burial deposit. The diameter varied but was usually 8–10 ft. The third type seemed to exist only in the Fars region of Iran. This type of cairn is in the form of a stepped structure, the number of steps being two or three. For example, at Bishezard[15] in Fars, one cairn of this type measured 34 ft. in diameter and had 3 tiers rising to a total height of 14½ ft. The second and third tiers had irregularly disposed recesses where the burial deposits were placed. According to Stein, this type of cairn might represent family crypts. At Sar-i-Asiab near Kirman, one course of large undressed stones was first laid in an outer ring. The space within was divided by alignments of smaller stones running both east-west and north-south. The remainder of the cairn consisted of irregularly piled stones which were small in the lower heap but bigger in the upper. One is not aware if this construction technique was repeated elsewhere. There is no evidence of this type in Pakistani Baluchistan. The fourth type also is not found in Pakistani Baluchistan. Its evidence comes from Asmangird[16] in Fars. The largest of the Asmangird cairns was about 38 ft. in diameter and 8 ft. high and revealed a roughly built passage about 6 ft. long leading towards the centre from the south-east and covered with four large slabs. The burial offerings lay in the centre.

The burials were in all cases fractional burials, though at least three sites—Fattehabad, Kalatuk and Zayak—revealed no burial remains.

[15] Ibid.

[16] Ibid.

This may also be due to the very limited nature of excavations at these sites. In a few cases, notably at Zangian, the bones bore traces of burning, thus suggesting an earlier cremation. In others there is no such sign and the logical guess is that the body was kept exposed before the bones were collected and put inside a cairn. Remains of food were found in four cairns—remnants of date in two, fish-bones in one and some undetermined stuff in the other. At Zangian two horse-heads were found in two cairns.

The primary pottery associated with these cairns is a coarse whitish pottery, red or yellow in colour with a white slip. Stein noted its occurrence from Fars to Zhob-Loralai. The pottery is wheelmade but in some cases there was possibly a handmade pottery which was crude and thus meant for the funerary purpose only. In the Zehlamban valley, Iran, a cairn revealed five complete handmade vessels. Some pottery fragments at Damban showed red and dark brown slip. Another type which was not uncommon was a green-glazed ware. A green-glazed bottle of elegant shape was recovered from Gatti while a number of green-glazed potsherds were present at Damba Koh. At Surjangal sherds of chalcolithic painted pottery were found inside the cairns but in this case the cairn-burial people living near the ruins of the earlier chalcolithic settlement at the site were obviously attracted by the available fragments of the earlier pottery on the surface.

From what has been published of the shapes of vessels, one can make out the following types: short-necked, ovoid, flat-bottomed or footed jars which are very often loop-handled and spouted, with perforated lugs around the body for suspension (in one particular example of this type, on one side of the loop-handle across the mouth, there is a pair of animals moulded in the round and standing on their hindlegs); open jars with pedestals; beakers, sometimes with a spout and perforated lugs; globular close-mouthed pots with a spout and a loop-handle; flat, ovoid, flask-shaped pots with a spout or what has been called a 'canteen-like vessel', etc. The decoration may take the form of mere incised wavy lines between sets of straight horizontal lines. In many cases, however, there is a painted surface with various geometric designs among which the hanging spirals seem to be a distinctive one.

The miscellaneous grave-goods comprised, in the main, the following objects: grinding stones, and stones for polishing knives, etc., beads of terracotta, glass, frit and stone, a decorated bone-piece, all pendants of blue frit, a glass stylus, a few pieces of shell ornaments and other objects, two small charms representing a frog and a lion couchant, a jade bead, fragments of glass, steatite and pottery spindle whorls, two wooden combs, a Parthian silver coin, a Parthian copper coin of Sinatruces (77–70 B.C.) and a Sassanian silver coin of Yazdagird III (A.D. 632–651). The metal objects were of copper-bronze, silver and iron. The copper-bronze objects consisted of a footed bronze platter from Damba Koh, two copper bracelets with a snake-head pattern, three copper bracelets without any pattern, a copper lamp or censer, a copper ring with a bezel and fragments of a copper vessel from Jiwanri, fragments of copper wire or pin from Take Dap, fragments of copper at Zangian, a bronze jar on tripod, three bronze 'cat-bells' and three bronze rings (one with a bezel carrying an intaglio design suggestive of the Kushana or Gupta art) from Mughal Ghundai, a bronze finger-ring from the Zehlamban valley and a bronze dagger-sheath with silver studs from Bishezard. There were some small thin silver rings at Jiwanri, one silver ring at Mughal-Ghundai, a portion of folded silver plaque, possibly part of a buckle, at Damban and a small silver ring at Bishezard. The iron objects were the following: a lump of oxide of iron, very thin iron vessels and two iron javelin heads from Damba Koh, some fragments of iron implements (unspecified) and an iron fish-hook from Jiwanri, fragments of an iron pot from Gatti, 'small pieces of iron implements' from Nasirabad, a broken large sword blade and a smaller one with a bronze fastening at the hilt from Zangian, ten arrowheads and one spearhead from Mughal Ghundai, four triangular spearheads and a fragment of horse-shoe from Bishezard.

It may be noted that copper-bronze implements like axe, chisel, etc. are completely absent at all the sites and so are stone implements. Though the typological range of the iron objects found is rather limited, there is no reason to suggest that the knowledge of iron as such was limited in any way. Besides, the excavations at the 'cairns' were unlikely to find a wide range of variations in objects.

The idea that the Baluchistan cairn-burials were prehistoric and associated with the beginning of iron in India was first put forward by D. H. Gordon in 1950.[17] He emphasized the continuous volute pattern of the cairn-burial painted pottery and then drew attention to the occurrence of this design on two small pieces of sherd from Chiga-kabud in the Alishtar plateau of central Zagros[18] and also to

[17] Gordon, 1950.
[18] Ibid.

some unpublished sherds with the same design from the site of Bagh-Limu in the Luristan region.[19] Chiga-kabud yielded two iron sickles with bent back tang, one of which was illustrated by Stein.[20] These belong to a type which also occurs in the Necropolis B assemblage at Sialk.[21] Gordon's argument was as follows: Chiga-kabud had a link with Sialk VIB because it had an iron sickle with bent back tang; Chiga-kabud also had links with the Baluchistan cairn-burials because the continuous volute pattern of the cairn-burial painted pottery occurred on two small pieces of sherd at Chiga-kabud. So the Baluchistan cairn-burials were culturally linked with Sialk VIB.

This seems to be Gordon's main argument. He tried to reinforce it by suggesting a few other analogies. Some objects from the Mughal Ghundai cairns—a copper bracelet of an adjustable type, a bronze jar on tripod, small bells (Stein called them 'cat-bells'), thin copper rings, and iron arrowheads—were supposed to have parallels among the finds from Sialk VIB. Among the arrowheads a three-flanged type also recalled a specimen from Nad-i-Ali[22] in Afghan Seistan and an analogy of the copper bracelet of the adjustable type was found at Giyan.[23] Attention was also drawn to the occurrence of the horse as an element of design on the Londo ware of Baluchistan[24] which, incidentally, was never found to be directly associated with the cairn-burials. The design element was supposed to have been derived from the horse friezes painted on some Sialk VIB pottery. In Gordon's argument, the Londo ware was contemporary with the cairn-burials (even though they were not found associated) and so, if the Londo ware was of Sialk VIB inspiration, the cairn-burials were similarly inspired too. In this context he drew attention to the presence of two horse-heads in two cairns at Zangian.

Once the Sialk VIB association of the cairn-burials was 'established', the problem of dating them became comparatively easy. There were only two issues involved: how to date Sialk VIB and how much time-gap to allow for the influence to spread to Baluchistan. In 1958 Gordon[25] accepted Claude Shaeffer's date of 1200–1000 B.C. for Sialk

[19] Ibid.
[20] See Chakrabarti, 1979.
[21] Ibid.
[22] Ibid.
[23] Ibid.
[24] Ibid.
[25] Gordon, 1958.

VIB[26] and put the beginning of the Baluchistan cairn-burials at around 900 B.C. He also gave the burial practice some time to spread from its beginning in the Makran area to Zhob-Loralai in the north-east and put the later group between 550 and 450 B.C.

What gives this chronological scheme of Gordon some amount of historical interest is that it came to be broadly accepted by so many scholars—W. A. Fairservis[27] (suggested date: 1000–800 B.C.), N. R. Banerjee[28] (suggested date: 800 B.C.) and the Allchins[29] (suggested date: 1100–750 B.C.) as well as others. The Sialk VIB association was taken for granted; it was not thought necessary to examine Gordon's basic premise in depth or detail. Banerjee, in fact, built on this dating his own scheme of the origin of south Indian megaliths, which tried to show that the Baluchistan cairn-burial people gradually pushed their way down to the south. Nobody asked why the issue of the dating of cairn-burials should be related to the issue of the dating of the Londo ware; after all, they were never found to be directly associated. Banerjee, on the contrary, paid a lot of attention to the Londo ware without even emphasizing the basic fact that what was known about the Londo ware was derived only from Beatrice de Cardi's surface collections.[30] Another fact which was ignored was that the distinctive Sialk VIB 'tea-pots' on which one primarily notices the Sialk horse friezes were totally absent among the Londo specimens. It was also thought unnecessary to mention that the pothook spirals, a distinctive Londo design, did not occur at Sialk.

The first criticism of this view of Gordon came from Leslie Alcock[31] in 1952. Apart from emphasizing the extremely tenuous and flimsy character of Gordon's argument for a Sialk association of the cairn-burials, he suggested some positive historical parallels from a much closer context for some Mughal Ghundai objects. The Mughal Ghundai bronze jar on tripod could be matched by the jar no. 207 in John Marshall's *Taxila*.[32] Two other bronze jars with tripods are illustrated from Taxila but typologically no. 207 comes closest to the Mughal Ghundai specimen. All these Taxila jars are from the Parthian level of the city. The so-called 'cat-bells' are common enough in

[26] Ibid.
[27] See Chakabarti, 1979.
[28] Ibid.
[29] Ibid.
[30] Ibid.
[31] Alcock, 1952.
[32] See Chakrabarti, 1979.

the cultural assemblage of Sirkap of the first century A.D.[33] Iron arrowheads, copper bracelets of an adjustable type and thin copper rings of Mughal Ghundai, for which analogies were sought by Gordon in the widely differing contexts of Sialk VIB, Nad-i-Ali and Giyan are not particularly distinctive types by themselves and thus of no comparative value, but if one seeks to find parallels, one can easily find them in Taxila.[34] The Mughal Ghundai copper finger ring showing a male with a high headdress carrying a spear and a bow, and a female in the long skirt suggested Kushan-Gupta workmanship to Stein. The design was not illustrated but the date is clear enough: between the first and the fourth-fifth centuries A.D. Alcock also hazarded a guess about the date of the Londo ware. He noted that the grog-backing of the Londo ware had a similarity with a later pottery around Quetta which was called by Stuart Piggott a Sassanian pottery.[35] His comment was: 'Though the percentage of grog is lower and the shape of articles differs from those in Londo ware, this Sassanian pottery may provide a clue for its dating.'

In 1964, Subrahmanyam[36] followed Alcock in the sense that he also called the Mughal Ghundai assemblage early historical, his main emphasis being on the copper signet ring and the pot with Classical festoons and medallions. In 1968, Lamberg-Karlovsky and Humphries[37] found no 'convincing parallels to the Sialk B culture of the material contents found in the cairn-burials'. On the other hand, they placed emphasis on the Taxila analogies of some of the Mughal Ghundai objects—the Mughal Ghundai pot with Classical designs, the Mughal Ghundai copper signet ring, the Parthian copper coin of Sinatruces (77–70 B.C.) at Damba Koh and the Sassanian silver coin of Yazdagird III (A.D. 632–651) at Bishezard in Fars. The open pedestal bowls from Fanuch and Suntsar could be matched by similar bowls from Taxila.[38] For the 'canteen-flask', a distinctive cairn-burial form, they found an analogy in the Susa Achaemenian village III[39] and also drew attention to an unpublished find of this 'typical Parthian form' from near Pasargadae. Their conclusion was that the evidence

[33] Ibid.
[34] Ibid.
[35] Ibid.
[36] Subrahmanyam, 1964.
[37] Lamberg-Karlovsky and Humphries, 1968.
[38] See Chakrabarti, 1979.
[39] Ibid.

suggested a date in the last two centuries B.C. but, at the other end, the cairn-burials could continue up to the early mediaeval period.

The evidence of a historical date for the cairn-burials is clear enough. The indisputable pieces of internal evidence are the following: a silver coin found at Damba Koh by Mockler, which was said to be a common Parthian type; a copper coin of the Parthian Sinatruces (77–70 B.C.) found at Damba Koh by Stein; a silver coin of the Sassanian Yazdagird III (A.D. 632–651) found at Bishezard in Fars; the small pot with Classical rosettes, festoons and medallions, and the copper signet ring from Mughal Ghundai. The Taxila analogies of some Mughal Ghundai finds like the bronze jar on tripod, etc. may not be particularly distinctive but in the light of the above-mentioned internal evidence they are far more convincing than any tenuous Sialk parallel. In fact, the Taxila analogies may be pushed a bit further. The flat, flask-shaped, lugged vessel which Mockler first illustrated from Damba Koh has an exact parallel in a find from Parthian Sirkap.[40] This is a distinctive shape and the importance of this parallel is obvious enough. It is also worthwhile to point out that the pothook spirals of the cairn-burial pottery occur quite significantly in the early centuries A.D. context in the Soviet Khorezm in central Asia.[41]

One may, in fact, safely assert that there is not a scrap of data at present, which would argue for any date beyond the first century B.C. for these cairn-burials. Those who are reluctant to find the remains of any early historical occupation in Baluchistan may only be reminded that Stein himself discovered a Buddhist stupa site at Tor Dherai[42] in the Zhob-Loralai area and that the Sampur mound, Mastung, Kalat, revealed a silver cup[43] which can be duplicated in all detail in Taxila.[44] It should also be emphasized that Stein[45] subsequently found at Sampur the coarse red ware with a white slip, a characteristic of the cairn-burial pottery. He also commented on a 'close similarity in shapes' between the pottery excavated by H. Hargreaves at the indisputably early historical site of Sampur and the pottery found by him at sites like Mughal Ghundai, Zangian and Jiwanri.[46] Finally, all

[40] Ibid.
[41] Ibid.
[42] Ibid.
[43] Ibid.
[44] Ibid.
[45] Ibid.
[46] Ibid.

controversies regarding the 'prehistoric' character of the Londo ware may be set at rest by pointing out that it has been found in the Sassanian context at Tepe Yahya.[47]

Table I
Chronological Range and Types of Iron Objects in Baluchistan

Chronology	*Site*	*Context*	*Type*	*Number*
c. 800 + B.C.	Pirak	Levels 1–6	winged arrowhead	several
c. 1st cent. B.C.- early centuries A.D.	Mughal Ghundai	cairns	arrowhead	10
			spearhead	1
c. 1st cent. B.C.-early centuries A.D.	Zangian	cairns	sword-blade	2
	Jiwanri	cairns	fish-hook	1
	Gatti	cairns	pot	fragments
	Nasirabad	cairns	indeterminate	indeterminate

2. The Northwest

Gandhara Grave Culture

The distribution of the Gandhara Grave culture is in the area north-northeast of Peshawar, bounded by the Kunar-Chitral on the west and the Indus on the east and extending up to Chitral in the north. The terrain is basically hilly but generally below 4000 m. The higher Himalayan massifs do not begin till Chitral. The sites tend to lie along the rivers of which there are quite a number including the Swat and her tributaries. The valleys are never wide, the space between the hill-slope and the river providing the most convenient spots for settlement. The cultivation was obviously done in narrow strips along the rivers just as it is done even now in this region. One cannot be sure of the extent to which the culture spread to the Peshawar plain and the Potwar plateau proper. Although the predominant impression is that it is a culture primarily confined to the hill valleys, there are indications that it spread out at some stage till Taxila in the Potwar plateau.[48]

It is not that the protohistoric occupation of the Swat valley and its adjacent area begins only with this grave complex. Giorgio Stacul's work at the cave site of Ghaligai has provided a convenient

[47] Ibid.
[48] Allchin, 1982.

archaeological index for the whole area.[49] What Stacul calls the 'archaic phase of the protohistoric graveyards of the Swat valley' is only Period V of the Ghaligai sequence. Period I, an assemblage of coarse handmade pottery, bone and stone tools and antlers and boar tusks, goes back to c. 3000 B.C. (calibrated). Iron appears in Period VII or the third and last phase of the grave complex. The beginning of iron was not accompanied by a significant break in the cultural sequence and so the cultural context of the iron-bearing horizon has to be appreciated to some extent against the background of the earlier copper-bronze one.

The distinctive type of the Gandhara graves had two pits—an oval or circular upper one excavated to a depth of 3–6 ft. and a rectangular lower one dug into the floor of the upper one and containing the grave remains. The rectangular lower pit was dressed on all sides by the locally available stone slabs and similarly covered. The upper pit was filled with earth and the area on the surface demarcated by a circle of stones. This general type had a few variations; for example, there were headstones in some graves but these were not present in others. In some cases the upper pit could be rectangular instead of oval or circular while the lower pit also could be oval in some cases without being rectangular. Some of the graves, particularly the children's, had only one pit, rectangular or oval or even circular, which was dressed and covered by stone blocks and slabs. The floors of some of the graves were found paved with stone. The graves usually lay aligned with the slope of the hills.

The burials were of three types—complete inhumation, interment of bones and ashes after cremation, and fractional burials. In the first type, the body lay on its side with the legs and hands drawn up. There could be single, double or more skeletons in one grave. At Timargarha, bodies of a male and a female have been found in one grave. In the second type, the cremated bones and ashes, usually the remains of more than one person, were put in box-like sarcophagi and urns. The urns had sometimes the shape of a human face—appliqué nose and holes for the ears and the mouth, and in one case even the incised eyebrows. The third type showed fractional burials of more than one person—sometimes adults of both sexes and children. In some cases the graves contained burials of two periods, the bones deposited earlier being swept aside to make room for the new deposit.

[49] Stacul, 1969.

The associated pottery included a grey ware of different shades and red and brown wares. The fabric was both coarse and fine. The coarse varieties were either handmade or turned on a slow wheel. The fine varieties were wheelmade. The shapes were wide-ranging, but principally globular vessels, cups-on-stand, pedestalled bowls, tall vessels, elongated vessels with corrugated body and spouted or beaked jars with handles. Each of these types had its own variations. The grave-goods, apart from those of copper and iron, included a spiral gold ear-ring, a few triconical silver beads, a glass pendant, beads, pins, awls, etc. of bone, terracotta spindle-whorls or beads and terracotta human figurines. Most of the copper objects were only different varieties of pin. Among the other types mention may be made of a harpoon, an arrowhead, a laurel leaf-shaped object, a knife blade, pendants, ear-rings, etc. The agricultural base was fairly diverse and included, among other things, the cultivation of rice.

As the number of excavated graves containing iron is still limited, it is not particularly easy to isolate features which may be said to be confined only to this period. The basic mode of burial of this period seems to have been fractional but the method of cremation was not entirely unknown. There may be certain changes in the pottery forms and style but none seems to be particularly striking.

The iron objects which have been reported from these graves may be put in the form of a list.

A List of Iron Objects of the Gandhara Grave Culture

Timargarha—grave 109—spearhead; stem with rectangular section; no midrib—1.

Timargarha—graves 192 and 112—disc-headed nail—2.

Timargarha—grave 142—cheek-bar of a horse's harness; rectangle-sectioned straight bar with three equidistant elliptical holes; a knob at the ends—1.

Timargarha—grave 149—spoon; rectangle-sectioned handle terminating in two rings, one on each side—1.

Balambat—Period III—finger-ring of rounded wire—1.

Katelai I and Loebanr I—graves (details not mentioned)—disc-headed pins or nails—27.

Katelai I and Loebanr I—graves (unspecified)—arrowhead; leaf-shaped and thickened in the middle; round-sectioned and pointed tang—2.

Katelai I and Loebanr I—graves (unspecified)—spearhead; leaf-

shaped and thickened in the middle; round-sectioned and pointed tang—7.

Katelai I and Loebanr I—graves (unspecified)—spearhead 'shaped like a very elongated leaf' and thickened in the middle; long, round-sectioned hollow tang—3.

Katelai I and Loebanr I—graves (unspecified)—axe; flat and rectangular; in one case with a rectangular projection from the body 2.

Katelai I and Loebanr I—grave (unspecified)—fork; flat and elongated; one end consists of four prongs—1.

Buner—unspecified grave—spearhead; leaf-shaped and thickened in the middle; small tang—1.

Noghurmuri—grave 1—arrowhead; two barbs and a tang—1.

Ghaligai—upper strata—pin (roundheaded); in one case with curved body—2.

Ghaligai—upper strata—unidentified object; triangular in section—1.

Ghaligai—upper strata—arrowhead; two barbs and a tang—1.

Ghaligai—upper strata—arrowhead; rhomboid in section—1.

The following basic types may be isolated among the total range of iron tools.[50]

1. *Spearhead* (Timargarha, Katelai I–Loebanr I, Buner)—total number—12. Two sub-types: (a) midribless specimen with a rectangle-sectioned tang from Timargarha and (b) leaf-shaped objects which are also midribless but thickened in the middle from the other sites. The latter group has two types of tang: pointed, and hollow, round-sectioned.
2. *Arrowhead* (Katelai I–Loebanr I, Noghurmuri, Ghaligai)—total number—5. Three sub-types: (a) leaf-shaped arrowhead with thickening in the middle, (b) arrowhead with two barbs and (c) arrowhead with rhomboid cross-section.
3. *Pin or nail* (Timargarha, Katelai I–Loebanr I, Ghaligai)—total number—31. Apparently two sub-types: (a) disc or round-headed, straight body and (b) round-headed, curved body.
4. *Spoon with a handle terminating in two rings, one on either side* (Timargarha)—total number—1.
5. *Finger-ring* (Balambat)—total number—1.
6. *Cheek-bar* (Timargarha)—total number—1.

[50] The basic sources are Antonini and Stacul, 1972, and Dani, 1967, but also Tucci, 1958, 1963; Antonini, 1969, 1973; stacul, 1966a, 1966b, 1967a, 1967b, 1969a, 1969b, 1969c, 1970a, 1970b, 1971, 1973a, 1973b.

7. *Unidentified object* but different from the other types (Ghaligai)—total number—1.

Among these types the cheek-bar from Timargarha deserves a special notice. In his study of the object, Karl Jettmar[51] points out that 'it belongs typologically to those groups which played a great role in the steppe-belt between the tenth and the sixth centuries B.C.'. His conclusion is:

> With attention to all the other objects in the Steppes, a dating in the seventh or sixth centuries B.C. could be tentatively proposed, and this agrees quite well with the dating of Period III as proposed by Dani (without however discouraging a later dating which Stacul evidently has in mind). The real importance of the piece lies in the fact that it once more points to the Steppes as one of the several sources of the Gandhara Grave Complex, and that it encourages us to look for other articles of horsegear in the material of sites from the subcontinent.[52]

There is some confusion about the date of the iron-bearing graves in the Gandhara Grave culture. According to Dani, these graves belong to his Period III, dated c. 900–c. 600 B.C.[53] According to G. Stacul, iron is associated primarily with his Period VII of the Ghaligai sequence. This dating makes this iron fall in the second half of the first millennium B.C. The problem has been expressed in the following fashion by the Allchins: 'The chronology of the graves, and of the early, middle and late phases proposed by Stacul, and somewhat differently by Dani, is still in need of clarification; but we are inclined to agree with Dani and place the late phase, in which it seems that the iron objects mainly occur, around the opening of the first millennium.'[54] Agrawal[55] thinks that 'the first appearance of iron in Swat area is dateable to c. 1000 B.C.'. In fact, we also believe that a date around the beginning of the first millennium B.C. in this case will satisfy the archaeological evidence.

Saraikhola

Period III at Saraikhola is represented by a number of graves belonging to two phases and lying atop the Kot-Diji-related occupation level at the site. Iron occurs in the second phase of these graves: 2 rings, 1 rod and a bracelet of paste beads, the ends of which are

[51] Jettmar, 1967.
[52] Ibid.
[53] Dani, 1967.
[54] Allchin and Allchin, 1982.
[55] Agrawal, 1982.

fastened with iron clasps. A date in the first half of the first millennium B.C. is possible.[56]

Balambat (Period IV), Bhir Mound, Charsada

These three sites are taken together because the iron objects from them roughly cover the second half of the first millennium B.C. Period IV at Balambat is supposed to be Achaemenid. The strata IV-II of the Bhir mound cover the pre-Mauryan and Mauryan levels, and the Charsada evidence includes the period between the end of the fourth and the second centuries B.C.

The objects from Balambat IV are: (1) 3 loop-headed pins (round-sectioned wire bent to form a loop); (2) 1 nail (square-sectioned stem and rectangular flat top); (3) 1 socket (square-sectioned and terminating in a point; ferrule of a staff ?); (4) 1 chisel (socketed tang); (5) 1 bar (rectangular section and a socket at one end); (6) 1 pair of scissors (called 'sheepshearer': flat bar hammered into two blades); (7) 1 knife-blade (rectangle-sectioned handle and straight-backed blade).[57]

Bhir Mound, stratum IV (sixth-fifth centuries B.C.) yields a dagger (double-edged, without cross-guard but with a tang; the blade terminates in a point). Its stratum III (fourth century B.C.) shows a scale-pan (a pan with rounded bottom with 2 loop-handles for suspension), a knife (straight-edged, straight-backed, with a tang) and an adze (thin, rounded top and thick, sharply tapering blade). The Stratum II (third century B.C.) has the following types: bowl (1), spoon (1), sieve (a fragment), dagger (1), javelin (three-flanged head; shaft and head in one piece; 1 in number), arrowhead (double-tanged; only in one case the blade is barbed and ribbed; the blades vary between lozenge, square and rhomboid in cross-section; 10 in number), elephant goad (1 in number; sharp point at one end and curved hook at the other); axe (socketed, drooping blade; 1 in number), adze (thin, rounded top and thick, sharply tapering blade; 1 in number), knife (straight-backed, straight-edged, tanged; 1 in number), tong (1), anvil (square in plan with a pointed leg at each corner; 1 in number) and nail (3 types—flat shank and head projected on one side and perforated at the upper edge from side to side; flat shank with splayed head, tapering to a point; circular or disc head with a tapering point; 10 in number). The stratum I of the Bhir mound

[56] Halim, 1970–1.
[57] Dani, 1967.

(third-second centuries B.C.) yields a spearhead (spike-shaped, foursided, tanged) and a spud (chiselheaded and socketed, the socket being formed by beating out the metal and bending over the two sides to meet in point).

The first iron-yielding level at Charsada (later fourth-beginning of the third century B.C.) showed a socketed spearhead, and its second century B.C. level yielded 3 arrowheads (2 specimens triangular and lozenge-shaped in cross-section, the third one leaf-shaped and quadrangular in cross-section) and one flesh-hook (two-pronged and socketed).[58]

Among the finds of a recent excavation[59] at the Bhir mound one notes at least 5 additional types in the Mauryan level: key, rings, clamp, staples, door-rings, and *kalpi* (a wide piece of iron with the upper end curved and pointed). This addition makes the total number of types from the Mauryan Bhir mound 18. What is also important is the quantity of slag found in the different levels of the site during the recent excavations.

Sirkap

Sirkap in Taxila has 7 strata. The earliest, the seventh one, goes back to the pre-Greek times when Sirkap was possibly a suburb of the main settlement of the Bhir mound. The sixth and fifth strata belong to the Indo-Greeks by whom the city was founded, and roughly cover up to the first decade of the first century B.C. from the closing years of the third century B.C. when the city was possibly founded. The fourth, third, and second strata all belong to what has been called the Saka-Parthian period and take the history of the site to the middle of the first century A.D. or slightly earlier. Due to the confused chronology of Indian history of the period one cannot be absolutely certain of the different points of date but the broad outline of the Sirkap chronology is clear enough.

Marshall listed the different types of iron objects from Sirkap, grouped into seven broad categories: household utensils, arms and armour, horse-bridles and elephant goads, carpenters' and blacksmiths' tools, agricultural implements and miscellaneous pieces. Of these, most of them are found in the strata III–II or the late Saka-Parthian levels, dated roughly in the first century B.C.–first century A.D. There is no iron object in the stratum VII; there is only one class (chisel) of object in the strata VI–V and there are only four classes of

[58] Wheelar, 1962.
[59] Sharif, 1969.

objects (spoon and ladle, scale pan?, arrowhead and miscellaneous) in the stratum IV. In this context it is worth recalling John Marshall's caution that only a small fraction of the total number of iron objects discovered at Taxila has been reported, and that the objects stood a better chance of survival in the generally burnt-out, ashy deposits of the late Saka-Parthian strata.

The types of iron objects found in the strata III–II (the late Saka-Parthian levels) of Sirkap are as follows.[60]

1. *Cooking pot or cauldron*: 4 specimens; rounded bottom, everted rim, 2 rings on the shoulder; made in two sections and either riveted or welded around the middle; height 7.75–12 ins.
2. *Tripod stand for pots*: 2 specimens; a ring supported by three legs, looped projections on the inside of the legs; diameter 10.62–10.75 ins.
3. *Bowl*: 3 specimens; round bottom and sides, lip sometimes incurved; diameter 4.62–4.75 ins.
4. *Dish and saucer*: 3 specimens; splayed sides, occasionally horizontal rim or slightly convex base; diameter 4.62–14.75 ins.
5. *Frying pan*: 2 specimens; in one, the pan has a concave depression at the middle surrounded by a broad horizontal rim and the loop handles are riveted to the edge; in another, two pans with round bottom and splayed sides are stuck together and each has a loop-handle riveted to the sides; diameter 13.5–22.25 ins.
6. *Spoon and ladle*: 4 specimens; the bowls of the ladles are spherical with a vertical handle; diameter 1.62–3.12 ins.
7. *Scale-pan ?*: 1 specimen; pan with 2 loop-handles for suspension; diameter 4.25 ins.
8. *Lamp*: 1 specimen; shallow bowl with pinched mouth; diameter 3.75 ins.
9. *Candelabrum*: 3 specimens; plain vertical rod with tripod base and terminal, occasionally with pairs of side-brackets; height 21–27 ins.
10. *Incense-burner*: 1 specimen; on three legs with a handle; diameter 3 ins.
11. *Wheeled brazier*: 2 specimens; rectangular and with wheels; corner-handles and with 2 rings attached to the body; 21 by 19.05 ins. (1 specimen).
12. *Shovel*: 2 specimens; flat blade and long, straight handle; length 26–26.40 ins.
13. *Bell*: 5 specimens; cylindrical or bowl-shaped with ring handle; diameter 2.5–8.5 ins.

[60] Marshall, 1951.

14. *Lock, key and lock-plate*: 1 specimen of key with ring-handle and teeth; 1 specimen of rectangular lock-plate with key-hole near the middle and holes for nails at the four corners; length 4.5 ins.
15. *Folding chair*: 1 specimen; hoof feet; height 26 ins.
16. *Sword*: 3 specimens; straight, double-edged with a cross-guard; length 21–34.25 ins.
17. *Dagger*: 4 specimens; double-edged, straight-bladed, with cross-guard which in one case was shod with bronze, and with tang; length 7.75–11.5 ins.
18. *Spear*: 2 specimens; one leaf-shaped and the other dagger-shaped, both socketed; length 7.12–8.25 ins.
19. *Javelin*: 2 specimens; four-flanged head with long shaft; length (broken) 22–25 ins.
20. *Arrowhead*: 3 specimens; knife-blade head, conical head with circular cross-section and three-bladed head; length 2.12–4.1 ins.
21. *Armour, helmet and shield-boss*: 18 armour plates and 3 links of an iron chain found much corroded and adhering together; the plates of different sizes and patterns, and curved to fit the body; edges straight or curvilineal or provided with hinges or pierced with small holes for lacing; each made of separate strips of metal either hammered or riveted together; the size and weight of the plates suggest use as armour for horses or camels; 2 groups of plate-armour intended for the arms; 1 helmet with cheek-piece on one side (the other side missing); its crown is of one piece beaten out like an oval bowl and afterwards deepened by means of a horizontal band hammered on to it; on the summit is a boss intended for a ring, spike or crest; size 9.5 by 7 ins. 2 shields, one with single-looped cross-piece riveted at the back and another with 3-looped cross-piece; diameter 1.87–2.87 ins.
22. *Horse-bridle*: more than 4 specimens; simple, plain, jointed snaffle-bit, and the protected snaffle with a ring or cheek-bar at each side to prevent the rein from slipping into the mouth; length 5.5–8.62 ins.
23. *Elephant-goad*: 1 specimen; sharp point at one end and curved hook.
24. *Axe*: 7 specimens; 3 types, all socketed; the first type with a wide splay and thin blade; the second type with a blade prolonged in both directions so as to give a longer cutting edge, the third type with the socket projected up and down the back of the handle in order to meet the strain imposed on it; length 3.68–6.87 ins.
25. *Adze*: 3 specimens; the blade broad below, thick above and squared off at the top for insertion in a metal sheath; length 5.25–6 ins.

26. *Chisel*: 2 (?) specimens; 2 types; one is a round bar chisel with double slope and straight cutting edge, possibly for stone cutting; the other is a round bar carpenter's chisel tapering from the top and with broad, crescentic cutting edge; length 5.75–7 ins.
27. *Knife and chopper*: 2 specimens of knife; straight-backed, tanged and with a convex edge; length 4.75–5 ins; 1 fragment of a chopper.
28. *Saw*: 1 specimen; straight-backed, straight-edged; a fragment.
29. *Tong, plier and tweezer*: a number of specimens; length 3.5–10.25 ins.
30. *Scissor*: 1 specimen; a fragment.
31. *Hammer, adze-hammer, pick*: a number of specimens; square, flat-faced hammer, larger round-faced hammer, small adze-hammer and the single and double picks; length 3.37–6.12 ins.
32. *Anvil or beak iron*: 3 specimens; small anvil of stool-type, anvil of solid bar type, square in section with splayed top; size 2.75–3 ins. square.
33. *Nail, nail-boss and hook*: a number of specimens; heavy boss heads, ornamental door bosses; length 2.75–4.5 ins.
34. *Clamp and staple*: a number of specimens; clamp of straight flat strip of metal, pierced by nails; double clamping iron with single nail at one end only; also staples.
35. *Chain*: a number of specimens; figure-of-eight links.
36. *Spade, spud, hoe*: a number of specimens; socketed hoe with narrow or broad blade, socketed spud, spade with double rings for socketing handle and 2 rings for attachment of cord; size—hoe: 7.37 ins. in length, spade: 8.12 by 6.37 ins.
37. *Weeding fork*: a number of specimens; fork with a tang; length 5.25–5.62 ins.
38. *Sickle*: more than one specimen; one type with a curved blade and straight handle; another type with a curved handle and straight blades.
39. *Ingot*: more than a hundred specimens; long torpedo or shuttle-shaped pieces, hexagonal in section and measuring between 4.75 and 7.12 ins. in length and between 1.3 and 2.5 ins. in width at the middle.
40. *Miscellaneous*: rather unidentifiable types but some of them may be plummets, crow-bars, etc.

Shaikhan Dheri

Shaikhan Dheri in Charsada is dated between the mid-second century B.C. and the end of the second century A.D. Only a small number of objects have been reported. The greatest number of finds is said to be from what is called the House of Naradakha in the Kushan level and consists mostly of various types of nails, door-knobs, door-hooks, door-chains, door-shutters, timber joiners and various types of keys. The published tool-types are found also in Sirkap and thus need no detailed discussion. Three-flanged arrowheads are found in the post-Greek level and continue to occur in the Kushan level. A type of key has a splayed head.[61]

Dani draws attention to a major find in the following fashion: 'One important find from the same house (i.e., the House of Naradakha), trench C4 (2), consisted of a group of scales, each of oval shape, about 3 ins. in length, having a series of holes on the margin for fixing into an iron armour.' Allchin has analysed the specimen and he concludes:

> From what has been said it emerges that there was much in common in the methods and material of manufacture of armour in Parthia and Gandhara during the early centuries A.D. But there was considerable variation in the forms of armour manufactured and in their function. Should it be accepted that our piece was a helmet, it does not necessarily follow that it was native Gandharan work, since the context of its discovery, with fire and loot, is suggestive of attack from without, and it is quite possible that both the helmet and sword were dropped by a foreign invader.[62]

Later Sites

A wide range of sites of later periods reveals iron objects which are in any case briefly reported and do not deserve any particular mention.[63] One site, however, has a smithy in its Period V, dated sixth-tenth centuries A.D. and, as an excavated smithy is rare in all periods of Indian archaeology, it deserves special mention. Abdur Rahman, the excavator of the site, has divided the iron implements from the site into four groups: (1) blacksmith's and carpenter's tools, (2) agricultural implements, (3) arms and (4) miscellaneous. The first group comprises chisels, fire-tongs, hammers and anvil. With two exceptions, all of them come from the smithy which measured 18 ft. by 13.5 ft in area. The kit of tools was found *in situ*. Found in association with this floor-level was an open-mouthed bowl probably meant for keeping water for cooling down metal objects after they were heated

[61] Dani, 1965–6.
[62] Allchin, 1970.
[63] cf. Dar, 1970–1.

and hammered into shape. One could also trace the furnace in this context. The agricultural tools comprised sickles and tools for breaking ground and weeding purposes. The second type was unsocketed with a narrow chisel-like blade and a curved tang for wooden grip. The large number of arrowheads at the site were long, thin, square-sectioned and with a small tang intended for insertion into wooden or reed shaft.[64]

NUMBER OF TOOL-TYPES OF THE MAJOR EARLY IRON-BEARING PHASES IN THE NORTHWEST

Prehistoric—Gandhara Graves and related sites; Saraikhola—8.
The Achaemenid phase (sixth–fifth centuries B.C.)—Balambat IV, Bhir Mound IV—8.
The Mauryan phase (third century B.C.)—Bhir Mound II—18.
The Saka-Parthian-Kushan phase (first century B.C.–second century A.D.)—Sirkap III–II, Shaikhan Dheri—40.

Several of these broad types have their own sub-types. It is obvious that the tool-types of the first two occupational phases were comparatively limited. At the same time it is pointless to suggest that this reflects a limited use of iron in these phases. The tool-types, when examined in detail, are found to be reasonably diverse. Besides, most of the sites of the first phase are grave-sites; one need not expect a lot of household iron objects in them. Again, the evidence of only one site may be misleading. For instance, the Achaemenid level of the Bhir mound showed only one object, but the same level at Balambat showed seven types. The use of iron obviously increased in the Mauryan and Saka-Parthian-Kushan phases but this need not suggest anything more than an increased complexity of the economic and social life in these periods. It is also worth emphasis that no copper-bronze implement suggestive of a productive use has been found even in the first phases.

3. The Panjab Plains and Sind

The northwest may be said to continue on the east up to the Salt range, and the Panjab plains proper lie between the Salt range and the Sutlej. No early iron-bearing site has yet been reported in this area. In fact, the only site one can mention is Tulamba[65] but iron objects in any quantity seem to be reported only in its period III, dated eighth–twelfth centuries A.D.

[64] Rahman, 1968–9.
[65] Mughal, 1967.

Sind also does not have any known early iron-bearing site. There is a significant gap in the archaeological record of Sind between the end of the chalcolithic and the beginning of the early historic periods. The first archaeological evidence of early historic occupation in Sind does not antedate Brahminabad of c. first century B.C., but even then the evidence is meagre. As far as the data on early Indian iron are concerned, the Panjab plains and Sind do not have any significance in the present stage of knowledge.

4. The Indo-Gangetic Divide and the Upper Gangetic Valley

The broad geographical limits of this area are between the Sutlej on the west and the confluence of the Ganges and the Yamuna on the east. The first demonstrable evidence of iron comes from what is called the Painted Grey Ware level, the third in the regional post-Harappan sequence, the first two levels being called successively the Ochre Coloured Pottery and the Black-and-Red Ware levels.

Painted Grey Ware Phase

The Painted Grey Ware settlements are village settlements with wattle-and-daub houses and a subsistence pattern based on the cultivation of rice, wheat, barley and pulses and the use of domestic cattle, sheep, goat, buffalo, pig, horse, etc. A number of bone and copper implements occur with iron. The bone implements are mostly points and the copper ones are arrowheads, chisels, hooks, nailparers and antimony rods. The grey ware itself is a distinctive ware occurring only in a limited quantity. Its shapes are mostly straight-rimmed bowls, dishes with various geometric designs like swastikas, sigmas, short spirals, dots-and-dashes, etc. There were three main types of associated pottery—black-slipped, black-and-red and red (with or without a slip).

For both this phase and the subsequent early historic Northern Black Polished Ware level, Atranjikhera must be considered the most important site because the data from this site have been fully published. The credit of demonstrating the significance of this phase in the upper Gangetic valley archaeology goes, to B. B. Lal whose work at Hastinapur first traced the antiquity of iron in the Gangetic valley before the historic period.

The iron objects from the Painted Grey Ware phase at Atranjikhera have been divided by its excavator, R. C. Gaur, into three major groups—(i) weapons, (ii) craft tools and implements and (iii) household objects and ornaments. He further observes:

> The presence of iron objects in such profusion and the discovery of furnaces, slags and certain specific tools used by blacksmiths, suggest that not only were iron goods manufactured at the site, but that the smelting of iron ore was also carried out here. The main source of iron for the P. G. W. people of Atranjikhera was most probably the region extending from south of Agra to Gwalior containing rocks in which iron content is quite high.[66]

The main types of objects found in the Painted Grey Ware level of Atranjikhera are the following:
Arrowheads: 21 specimens in all, 7 from the middle phase and 14 from the upper phase; both socketed and tanged; mainly triangular and leaf-shaped bodies but there are variations; triangular blades usually show a midrib running from shaft to blade; they are also barbed in some cases; length 4.3 to 10 cm.
Spearheads: 8 specimens (3 from the middle, 5 from the upper); dagger-shaped long blade with a rounded blunt tip and conical tang for fixing it into a wooden shaft; length of the described specimen 16.75 cm.
Shafts: 10 specimens (lower–2, middle–5, upper–3), all socketed and uniformly circular in section; 'most of them appear to have been fixed with arrowheads'; length of the described specimen 9.7 cm.
Tongs: only a single pair (upper phase); length 39.5 cm. According to the excavator, 'its short width, thickness and particularly its tapering arms to semi-pointed ends, suggest that it served the purpose of a tool of the blacksmiths rather than a kitchen utensil. The tool is made of a long flat bar folded in two equal arms with a pronounced head, without any joint.'[67]
Clamps: 21 specimens (middle–10, upper–11); the common shape is a straight band of iron with one or two holes at either end for nailing; the maximum size among the three illustrated specimens in the report: 11.2 by 2.5 cm.
Chisels: 6 specimens (middle–4, upper–2); divided into three types: roughly round head, flat bar with tapering sides and roughly flat square head; maximum length 12.5 cm.
Bars or rods: 7 specimens (middle–2, upper–5); rectangular, flat and circular in section; maximum length 17.5 cm.
Borers: 6 specimens (lower–1, middle–2, upper–3); straight circular bar with pointed working and rounded top; maximum length 14.7 cm.
Needle: only one specimen (upper phase); flat long bar with slightly

[66] Gaur, 1983: 219–20.
[67] Ibid: 224.

tapering sides and a prominent eye for the thread; length 9.8 cm. 'Its size suggests that it might have been used for leather stitching.'[68]

Hooks: 7 specimens (middle–2, upper–5); 'the hooks were evidently intended to be fixed in the walls at convenient places for use as hangers': long, straight hook bent at one end with tapering pointed tip and rectangular cross-section; maximum length 16.1 cm.

Nails: 20 specimens (lower–2, middle–4, upper–20); knob-headed, hook-headed, flat squarish head, etc.; maximum length 7 cm.

Axe: only one specimen (upper phase); straight butt and slightly splayed edge.

Knife: 3 specimens (middle–1, upper–2); straight back and convex edge.

Bangle: 2 specimens (middle–1, upper–1); broken.

Slag: 7 pieces (lower–1, middle–2, upper–4).

Indeterminate fragments: 14 specimens (lower–2, middle–3, upper–9).

The middle and upper phases of the Painted Grey Ware level at Hastinapur yielded knife-blade, nail, arrowhead and slag. Arrowhead, spearhead and nail occurred in the Painted Grey Ware level at Alamgirpur. The recently excavated site of Hulas in the Saharanpur district is said to have yielded an impressive quantity of slag. Period II of Jakhera has been divided into two sub-periods—A and B. In period IIA of Jakhera which has only 'a few pieces of PGW' from the upper deposits, there are reports of iron slag and bloom. Iron objects associated with Period IIB, supposedly representing 'the mature PGW culture' include hoe, sickle, spearhead, arrowhead, dagger/chopper, chisel, axe, nails, rods, etc. There is no doubt that iron was a widely used material in the Painted Grey Ware level of the Indo-Gangetic divide and the upper Gangetic valley. Could this be introduced in the earlier black-and-red ware deposit? One would need more data to come to a positive conclusion, but one notes the occurrence of shapeless iron fragments in the black-and-red ware deposit at Noh and also the presence of iron slag and bloom in the period IIA of Jakhera, which contains Painted Grey Ware only in its upper level.[69]

The chronology of the Painted Grey Ware level is still somewhat uncertain. B. B. Lal who first isolated the culture in a stratigraphical sequence at Hastinapur bracketed it between 1100 and 800 B.C. at that site. The only more or less fixed line Lal could depend on was

[68] Ibid: 228.

[69] Sahi, 1978.

the date of the Northern Black Polished Ware in the succeeding cultural level. This particular ware is generally synonymous with the beginning of the early historic period in the Gangetic valley and is dated accordingly around 600 B.C. At Hastinapur there was a significant break of occupation between the Painted Grey and the Northern Black Polished wares, which, according to Lal, could cover about 200 years. Assuming that the break was for about 200 years, the end of the Painted Grey Ware at the site could be put at around 800 B.C. The total occupational deposit of the Painted Grey Ware was about 7 ft. and Lal suggested that it could take 300 years to accumulate. So the beginning of the Painted Grey Ware at the site was placed around c. 1100 B.C. This chronological scheme of Lal was obviously based on two premises both of which were subjective and thus open to criticism, but there is no reason to believe that the critics like D. H. Gordon and R. E. M. Wheeler who tried to bring the chronology down to 800 or even 650 B.C. were in any way more logical than Lal. They differed from Lal basically in their estimate of the date of the beginning of the Northern Black Polished Ware which, according to them, was considerably later than 600 B.C.

What has cast significant doubts on the date proposed by Lal is the general range of C–14 dates obtained from a number of Painted Grey Ware levels, apart from the fact that the transition from the Painted Grey Ware phase to the Northern Black Polished Ware phase is now found to be continuous. All radiocarbon dates except TF-191 from Atranjikhera are later than 1000 B.C. The uncalibrated version of TF-191 is 1025 ± 110 B.C., and its consensus calibration has been calculated to be 1265–1000 B.C.[70] As D. P. Agrawal puts it, 'except for TF-191, all the other dates provide a bracket of c. 800–350 B.C. for the PGW.'[71] The weight of the radiocarbon evidence is no doubt in his favour but still one would like to point out that this comparatively low chronology does not conform to the general position of the Painted Grey Ware in the stratigraphical sequence of the upper Gangetic valley. There are several thermoluminiscent dates for the Ochre Coloured Pottery level, the first level in the sequence, at Atranjikhera and a few other places.[72] The dates show a wide scatter, but even on a conservative estimate, a date in the first half of the second millennium B.C. is suggested. The succeeding black-and-red ware

[70] Possehl, 1988.
[71] Agrawal, 1982.
[72] Ibid.

level in the region does not appear to be of long duration. The point is that if one has to put absolute reliance on 800 B.C. as the date of the beginning of the Painted Grey Ware, one has to admit that the Ochre Coloured Pottery and Black-and-Red Ware levels at Atranjikhera and other places in the *Doab* account for at least 700 years of occupation in this region.

Considering the occupational remains of these two levels it is a singularly improbable hypothesis. The theoretical possibility of a date around 1000 B.C. for the early phase of the Painted Grey Ware cannot be denied. In fact, if the thermoluminiscent dates of the Ochre Coloured Pottery level are acceptable, the date of the beginning of the Painted Grey Ware should be still earlier. We feel that the date of 1100 B.C., initially suggested by B. B. Lal, is an acceptable date in this context.

Incidentally, an interesting piece of evidence comes from the Painted Grey Ware level at Jodhpura in the eastern part of Rajasthan. On the authority of R. C. Agrawal and V. Kumar, the excavators of Jodhpura, D.P. Agrawal refers to the discovery of a 'crucible-shaped furnace used for direct reduction of ore', where the bloom was heated in an open furnace and forged on an adjacent platform. This type of furnace is a bowl-shaped cavity in the ground and among the most primitive types of furnaces.

In the easternmost section of the main distribution area of the Painted Grey Ware, Kausambi yielded in its Painted Grey Ware level only shapeless fragments of iron. This seems to be the general nature of evidence from this level at Ahichchhatra in the *Doab* as well. However, this does not in any way indicate that there was any dearth of iron in the Painted Grey Ware phase of the *Doab*.

Early Historic Period

The beginning of the early historical period in the Gangetic valley as a whole is generally marked by the appearance of the Northern Black Polished Ware and it is broadly dated between c. 600 and c. 200 B.C. with the possibility of its being earlier by at least a hundred years, as the dates from the site of Sringaverapura now indicate. While trying to tabulate the early historic iron objects from the Indo-Gangetic divide and the upper Gangetic valley our attention is dominated again by the fully reported plethora of finds from Atranjikhera. However, before discussing the material of this period from Atranjikhera, one may refer to the objects from the early historic levels at Rupar, Hastinapur and Kausambi which happen to be major

sites in the archaeology of the region. The basic iron tool-types from the early historic level of these three sites are as follows:
Rupar: (1) dagger, (2) knife, (3) nail, (4) hook, (5) bar, (6) spike, (7) handle, (8) sickle, (9) spearhead.[73]
Hastinapur: between c. 600 B.C. and c. 200 B.C. the types are: (1) chisel, (2) arrowhead, (3) bracelet (?) and (4) sickle; and between c. 200 B.C. and c. A.D. 300 the types comprised (1) nails of 3 types (knob-head, round section with bent head, and square section with bent head having a hole), (2) long chisel (?), (3) miniature bell, (4) adze, (5) base of a pan, (6) tube, (7) spike (?), (8) sickle and (9) chisel (?).[74]
Kausambi (c. 600–c. 45 B.C.): (1) arrowhead (11 types based on cross-section, tang, barb and blade), (2) spear, (3) javelin (spear and javelin of 5 types based on cross-section, blade and tang), (4) axe, (5) adze, (6) knife, (7) chisel, (8) sickle, (9) nail, (10) ring.[75] About 1,115 iron objects were recovered from Kausambi of this period in the 1957–9 excavations alone.

At Atranjikhera the early historic period or the Northern Black Polished Ware culture has been divided into four phases—A, B, C and D. The phase-wise distribution of iron objects (*after R. C. Gaur*),[76] can be seen in the table on P. 64.

Nine types of arrowheads have been illustrated by Gaur from this period. The classification is dependent on the minor variations in the shape, form and cross-section of the blade and the shaft. The length is roughly from 4 to 10 cm. The spearheads belong to seven types, the main variation being between elongated thick blades and leaf-shaped flat blades. The length goes up to 26 cm. The pieces of shafts which have been found from this level belong to spears and arrows. The agricultural implements comprise sickles, spud (prototype of modern spade), ploughshare, hoe and *khurpi* (described by Gaur as 'digger'). The ploughshare is easily identifiable: 'elongated share with tapering sides and corroded pointed end. It has a deep depression in the upper half portion for fixing it to the wooden frame of the plough.'[77] The objects which have been put together as 'craft tools and implements' are tongs, clamps, ring-fasteners, socketed clamps, staples, bolts, plumb-bobs, nails, hooks, chopping knives, choppers, pipes, scrapers, chisels (basic shape: slightly tapered with double

[73] Y. D. Sharma, 1953.
[74] B. B. Lal, 1954–5.
[75] G. R. Sharma, 1960.
[76] Gaur, 1983.
[77] Ibid.

		A	B	C	D	Total
1. Arrowhead	—	10	9	8	3	30
2. Spear-head	—	16	6	1	1	24
3. Shaft	—	3	1	1	-	5
4. Sickle	—	2	-	1	1	4
5. Spud	—	1	-	-	-	1
6. Ploughshare	—	-	1	-	-	1
7. Hoe	—	1	-	-	-	1
8. *Khurpi*—	-	1	-	-	1	
9. Tongs	—	-	1	-	-	1
10. Clamp	—	29	4	2	4	39
11. Ring-fastener	—	1	2	1	-	4
12. Socketed clamp	—	1	2	-	1	4
13. Staple	—	-	1	1	1	3
14. Bolt	—	-	-	1	-	1
15. Plumb-bob	—	-	-	3	1	4
16. Nail	—	22	13	17	21	73
17. Bar	—	6	2	2	2	12
18. Hook	—	11	3	1	3	18
19. Borer	—	13	5	-	-	18
20. Chopping knife	—	1	-	-	-	1
21. Chopper	—	-	-	10	-	10
22. Pipe	—	-	2	1	-	3
23. Scraper	—	-	2	1	-	3
24. Chisel	—	11	2	1	-	14
25. Axe	—	1	-	-	-	1
26. Knife	—	6	3	3	1	13
27. Lid	—	-	-	-	-	1
28. Disc	—	-	1	-	-	1
29. Bangle	—	2	1	-	-	3
30. Lump (slag)	—	1	4	6	4	15
31. Indeterminate	—	9	13	8	7	37
Total		147	79	70	50	346

slope and straight cutting edge) and axes. The household objects comprised knives (both curved and straight blades), bangles, lids and discs. Atranjikhera, in fact, offers us the most detailed inventory of iron objects from the early historic level in the upper Gangetic valley.

5. The Middle Gangetic Valley and the Adjacent Areas

The situation in eastern U.P. is not quite clear, and as far as I understand it, the black-slipped ware phase may denote the earliest iron-using phase in the central section of the Gangetic plain (cf. the black-slipped ware level at Ganwaria in Basti district, U.P.). The evidence from two sites, both in Allahabad district, appears to be of considerable interest. At Koldihwa the iron-yielding level includes iron axes and arrowheads, besides crucibles and slag. This level is said to be a continuation of the earlier chalcolithic level. 'Except for iron, other material equipment of this period is same as that of preceding cultures.'[78] The second site is Panchoh.[79] Apparently a single culture site, this showed three layers from the top: a 20 cm thick whitish layer, a 25 cm thick blackish layer with small stone pieces and iron nodules and a 15 cm thick yellowish layer with iron nodules and *kankar*. The pottery was represented by handmade and ill-fired corded and plain red wares. The other items comprised, among other things, small neolithic celts and microliths.

Along the Ganges in eastern India, the earliest iron-using level is marked by a black-and-red ware which appears in the earlier chalcolithic period. The pattern is the same at all the relevant sites—Sonpur, Chirand, Taradih, Pandu Rajar Dhibi, Mahisdal, Bharatpur, Bahiri, etc. The Iron Age evidence is not equally clear at all these sites but Chirand, Pandu Rajar Dhibi, Mahisdal and Bahiri deserve special mention. The black-and-red ware chalcolithic level at virtually all these sites falls in the second millennium B.C., according to the calibration chronology. However, at Bharatpur on the bank of the Damodar in Burdwan district, West Bengal, the beginning seems to be around c. 1700 B.C. This suggests that the major sites like Chirand near the confluence of the Gandak and the Ganges in Bihar might have known the beginning of the chalcolithic level as early as c. 2000 B.C. What is particularly interesting is that, even after the beginning of iron at these sites, the earlier elements, mostly the black-and-red ware and the microliths, continue to occur in a significant quantity. In fact, apart from iron there is very little to distinguish the iron-bearing phase from the earlier chalcolithic one. The radiocarbon dates (calibrated) suggest c. 1000 + B.C. as the general date of the beginning of iron in this region. Many dates fall in the first quarter of the first millenium B.C.

An interesting situation has emerged at a black-and-red ware site called Bahiri in Birbhum district, West Bengal.[80] First, in the upper level of the black-and-red ware phase there is an extensive deposit of iron ore and slag. In one area this deposit could be traced for about

[78] *IAR* 1973–4: 27.
[79] *IAR* 1975–6: 47.
[80] Chakrabarti and Hasan, 1982.

20 m. The fact that an area of 20 m across is full of iron slag and ore is perhaps worth pondering over. The site is on an edge of the Ajay flood-plain but, about 1 km away, there is an extensive lateritic spread locally known as *Muluker Danga*. There is no doubt that the iron ore utilized by the ancient settlers of Báhiri came from this lateritic spread. The wide occurrence of iron ore and slag in the upper levels of the black-and-red phase of the other excavated black-and-red ware sites in the western part of West Bengal, which is a continuation of the Chhotanagpur plateau, is a fact, as far as I am aware, although none of these other sites is as fully documented as Bahiri.

What gives the evidence from Bahiri its special interest is not the rich deposit of iron ore and slag in the upper level of its black-and-red ware phase but the occurrence of a limited number of pieces of iron slag above the natural soil in three exposed sections. The material is properly stratified. At the moment Bahiri stands isolated but the possibility that small pieces of iron slag in the 'chalcolithic' level were ignored in the excavated black-and-red ware sites in this region cannot be denied. An early calibrated date-range (PRL–866) for Bahiri is 1120–795 B.C.

This chronology is also supported by the dates from the so-called neolithic level of Barudih in Singhbhum district, Bihar, which is marked by a profusion of neolithic celts, coarse black-and-red pottery and an iron sickle. The earliest available date is 1055 ± 210 B.C. (calibrated 1410–830 B.C.).[81]

The nature of material found at some of these east Indian sites in their pre-Northern Black Polished Ware black-and-red ware phases will be clear from the following list.

Pandu Rajar Dhibi: spearhead, slag and oven. There was a row of elliptical ovens flanked by a thick layer of ash containing the spearhead and slag.

Mahisdal: arrowhead, spearhead, chisel, nail, abundance of ore and slag. I can testify that there was a profusion of microliths (made predominantly of crystal quartz) in this level.

Bahiri: iron slag in the lowest level of the deposit.

Barudih: an iron sickle. Slag occurred at this site too.

The details from the relevant sites of Bihar are not available but Chirand is known to have yielded 'many specimens' of iron. It may be emphasized that in both Assam and Orissa there is no evidence of iron before the beginning of the historic period in these areas. Orissa, however, has some black-and-red ware sites and, when excavated,

[81] Possehl, 1988.

these will surely throw light on the beginning of iron in this region. If the manufacture of big guns is any indication, Assam had an advanced tradition of iron metallurgy in the late mediaeval period. Apart from a primarily typological study of stone artefacts, the archaeology of Assam has not been worked out and it is impossible to pinpoint the beginning of iron in this region. Assam seems to enter the full light of history only in the Gupta-post-Gupta period but there is a possibility that the beginning of the historic period may go back to the third-second centuries B.C. No iron object has been recovered from the early period.

There is no dearth of iron objects from the excavated early historic levels in the middle Gangetic valley and the adjacent regions. A detailed site-wise inventory is not particularly edifying in the context of this region because the inventory from the early historic Atranjikhera is fairly representative of the general range of objects found in this region too. However, a reference to the Pataliputra excavations in 1951–5 may give a bare idea of the sources.[82] Period I at the site was dated before c. 150 B.C. and Period IV was dated A.D. c. 450–c. 600. Iron objects were found in large quantities in all the periods but most of them were corroded beyond recognition. Most of the published objects are nails of different varieties but some are other types like chisel, knife-blade, sickle, etc.

One would, however, like to draw specific attention to a site called Saradkel in the Ranchi area of the Chhotanagpur plateau.[83] The site has two occupational phases but belongs to the same culture which is dated first-second centuries A.D. The site is briefly published but the excavator referred to an 'astonishingly large number of iron objects' like chisels, arrowheads, axes with double or single cutting edges, longitudinal ploughshares, caltrops, door-hinges, rings, knives, etc. Another significant aspect was the occurrence of a number of pits which, from their content of charcoal, slag and sand, would appear to be iron-smelting ovens. In this connection one may also refer to the site of Karkhup in the Monghyr area, where a large number of pieces of iron slag and terracotta crucibles were found in an undated context.[84]

[82] Altekar and Misra, 1959.

[83] *IAR* 1964–5: 6.

[84] *IAR* 1960–1: 5. An 'ironsmith's workshop', dated c. 200 B.C., has been located at Khairadih in the Ballia district of U.P. in the middle Gangetic valley (*IAR* 1983–4: 86). This consisted of a number of furnaces 'built in close proximity to each other'. For a general discussion of the iron-bearing middle Gangetic valley sites, see Roy, 1983.

6. Rajasthan

In those areas of Rajasthan which border the Indo-Gangetic Divide and the upper Gangetic valley, the archaeological sequence is similar to that of these regions. From the point of view of the beginning of iron a major site is Noh near Bharatpur where, like the *Doab* site of Atranjikhera, the first two levels are successively of the Ochre Coloured Pottery and black-and-red ware. 'Shapeless' iron fragments have been reported from the black-and-red ware deposit. The third level is marked by the Painted Grey Ware. This level yields spearhead, leaf-shaped arrowhead with a socketed tang, and axe with a broad cutting edge. Arrowhead seems to be the only type to be reported from the succeeding historical level.

The most significant site in this context in Rajasthan is Ahar. It is the type-site of the chalcolithic Ahar or Banas culture of south-east Rajasthan and shows three phases of the chalcolithic occupation. Most interestingly, however, iron has been reported from the phases b and c of this 'chalcolithic' occupation: Phase b—1 arrowhead, 1 ring and slag; Phase c—4 arrowheads, 2 chisels, 1 nail, 1 peg and 1 socket. The radiocarbon dates put these phases of the 'chalcolithic' Ahar well within the second millennium B.C. In fact, the calibrated versions of these dates take them back to the first quarter of the second millennium B.C. These objects are spread in five trenches and belong to different layers, and by no stretch of the imagination can they be considered as belonging to disturbed occupational levels.

A major early historic site in Rajasthan is Bairat, although the scale of work at this site has been relatively minor. Bairat, Sambhar, Rairh and Nagari have all third century B.C. levels but their later deposits are not well dated. The iron objects belonging to these sites may, however, be broadly taken to belong to the early historic period. It should be noted that the present evidence is limited exclusively to east Rajasthan. A Kushan site, Rangmahal, in the Ghaggar valley of west Rajasthan is the only major excavated early historic site in that entire region. The evidence of iron in the Kushan levels of Rangmahal is not impressive.[85] The early historical levels of Bairat, Sambhar, Rairh and Nagari have revealed the following types of iron objects:
Bairat: clamp, nail, 'fish-plate', chisel.
Sambhar: axe, adze, mouthpiece of a blacksmith's bellow, staple,

[85] For Rajasthan, see Sankalia, Deo, Ansari, 1969; Sahi, 1979; Sahni n.d.a, n.d.b; Bhandarkar, 1929; Rydh, 1959.

'fish-plate', brace, ring, pivot, bell, ladle, spoon, arrowhead, spearhead, 'door-fitting'.
Rairh: sword, spearhead, dagger, knife, arrowhead, sickle, axe, adze, nail, 'door-fitting', chain.
Nagari: arrowhead, nail.

7. *Gujarat*

There is as yet no evidence of prehistoric iron in Gujarat. Even the evidence of early historic iron is scanty. The reason is that no early historic level in Gujarat has yet been extensively excavated and reported. Gujarat enters her historic period in the late fourth or early third century B.C. as it became a part of the Mauryan empire of Chandragupta I. It is not yet clear how much earlier, if at all, this beginning might have been. In the early historic level at Prabhas Patan (fourth-first centuries B.C.), which possesses a fortified complex, iron is reported.[86] Iron occurs in Period I of Nagara, dated between the fifth century B.C. and the beginning of the Christian era.[87] An iron-smelting industry has been reported at Dhatva,[88] which has been given a rather wide time-range, c. 500 B.C.–c. A.D. 200. The iron objects from the Buddhist site of Devnimori[89] belong to a period from about the third-fourth centuries A.D. to the seventh century A.D. or later. Out of 372 iron objects recovered, 240 were nails, which is not a surprising feature considering that they came mostly from the monastery complex and were possibly used for roofing tiles. The nails have been classified into six types: flat head, knobbed head, folded head, hood-like head, hammer head and featureless head. The rest of the objects comprised 3 arrowheads (socketed, and with triangular or barbed blades), 5 knives (3 types: straight-backed with tapering edge and pointed end, parallel-sided with rounded tip, concave back and upturned tip), 2 daggers, 2 rings, 1 chisel, 1 pick, 1 trowel, a pair of scissors, a door-chain and a hook, and a few other objects of indeterminate use. The daggers possess cross-guards but their blades are not ribbed. Period I at Timbarva in Gujarat, which was associated with the Northern Black Polished Ware, yielded arrowhead, fragment of a blade and possibly a ploughshare.[90] Iron objects have been described in the context of the

[86] *IAR* 1956–7.
[87] Mehta, 1968.
[88] *IAR* 1967–8.
[89] Mehta and Chowdhary, 1966.
[90] Mehta, 1955.

tenth century A.D. level at Dwaraka: fragment of an axe, knife, sickle, arrowhead, weight, etc.[91] Among the early historic sites of Gujarat, Nagara yielded a large number of iron objects—mostly nails (featureless heads, flat circular heads and hooked heads) but a few knives, spearheads, arrowheads, sickles, daggers, fish-hooks and some miscellaneous objects as well. The specimens, however, belong mostly to the late levels (eighth-ninth centuries A.D.) of the site.

8. Malwa

Iron in the Pre-Northern-Black-Polished-Ware Context in Malwa
The index sites of the beginning of iron in Malwa are Nagda and Eran, the former on the Chambal and the latter on the Bina. Both these sites show a remarkable continuity of occupation between the chalcolithic and the early historic periods and possess a substantial deposit of post-chalcolithic but pre-early-historic occupation having iron.

The basic components of the period I at Nagda are a red or cream coloured pottery with primarily geometric paintings in black, a profusion of microlithic tools and a limited use of copper. This ties up with what is called the Malwa culture typified at Navdatoli in the same geographical region. Period II is found separated from Period I by a thin layer of sterile soil which on analysis has been found to be the result of decomposition of vegetable matter in stagnant water. The site obviously was not deserted for a long period. The earlier pottery types do not disappear in the period II but the dominant ceramic now is a black-and-red ware. The use of the microliths continued but a new feature was the appearance of iron right in the lowest level of the deposit which is about 7 ft. thick. Period III was early historic and marked by the Northern Black Polished Ware. The sequence at Eran is exactly similar, though the periodization adopted is different. Period I is chalcolithic, also belonging to the Malwa culture. Period II has two sub-phases, A and B, B being early historic with the Northern Black Polished Ware. The Nagda-type black-and-red ware and iron occur in the period IIA. One may also refer to two other sites, Ujjain and Besnagar, to demonstrate the general stratigraphic position of iron in Malwa. Period II at Ujjain possesses the early historic Northern Black Polished Ware but a black-and-red ware occurs in Period I and, along with it, iron. The occupational deposit of this period is about 7 ft. thick. Ujjain had in this period a massive mud fortification, protected in places from the

[91] Ansari and Mate, 1966.

river Sipra by a reinforcement of well-cut wooden beams. An iron spade and a crow-bar have been found in the mud fortification giving an idea of the tools used. The types of wood used for the beams in the reinforcement—teak (*Tectona grandis*) and khair (*Acacia ferruginea*)—suggest also that iron tools must have been used to cut them. Besides, the habitational deposit of this period contains iron objects. Similarly, at Besnagar, the period I which is pre-early-historic possessed iron objects and black-and-red ware. An idea of the pre-Northern-Black-Polished-Ware iron objects in Malwa may be obtained from the following list:

Nagda: Period II (from the earliest level of the deposit onwards)—the total number of specimens: 59; tool-types: (1) double-edged dagger, (2) socket of an axe, (3) spoon, (4) axe with a broad cutting edge, (5) ring, (6) nail, (7) arrowhead with a biconical cross-section and a tang, (8) spearhead with oval or square cross-section, (9) knife (a wide variety of shapes), (10) sickle.

Eran: Period IIA—the details are not mentioned.

Ujjain: Period I (from the earliest level of the deposit onwards)—tool-types: (1) spearhead, (2) arrowhead, (3) knife, (4) crow-bar, (5) spade.

Besnagar: Period I—the details are not mentioned.[92]

Nagda does not have any radiocarbon date from any of its levels. But the basic point is that Nagda I belongs to the Malwa culture for which there is a large number of radiocarbon dates from the Malwa culture level of sites like Navdatoli, Eran, Kayatha, Dangwada and Inamgaon. At Navdatoli there are 8 C-14 dates all of which except one are internally consistent and place the culture between c. 1700–1600 B.C. and 1400–1300 B.C. (uncalibrated). The general range of the Malwa culture dates at the other sites is also more or less the same. Even at Inamgaon which is much further south than the Malwa heartland the relevant C-14 dates are: (TF–1001) 1565 ± 93 B.C.; (BS–263) 1459 ± 135 B.C.; (PRL–77) 1460 ± 113 B.C.; (TF–1000) 1375 ± 82 B.C.; (PRL–133) 1 375 ± 108 B.C.; (TF–924) 1370 ± 206 B.C.; (PRL–59) 1355 ± 113 B.C.; (BS–277) 1220 ± 105 B.C. When calibrated, most of these dates show a range of 1800/1700–1500/1400 B.C.[93] That the Malwa culture came to an end soon after the middle of the second millennium B.C. should no longer be in doubt. For

[92] For Nagda, Banerjee 1986; for Ujjain, Banerjee 1965; for Eran and Besnagar, *IAR* 1963–4: 16. Period IIA in Eran is said to contain iron in its lower levels which are pre-Northern Black Polished Ware.

[93] Possehl, 1988.

instance, the calibrated range of the three C–14 dates from the Malwa culture level at Kayatha is the following: (TF–398) 2015–1710 B.C. (TF–397) 1775–1560 B.C.; (TF–676) 1650–1350 B.C. There is no reason why such dates should not apply to Nagda as well.

The second point is that there cannot be any appreciable time-gap between Nagda I and Nagda II. That the temporary desertion of the site suggested by a thin sterile layer could not have been for long has been repeatedly emphasized by its excavator, N. R. Banerjee. Banerjee's own estimate of this time-gap is that it should not be more than 50 years. He assumed that the Malwa culture at Navdatoli came to an end around 800 B.C., and as regards Nagda, he wrote that 'the cultural homogeneity with Navdatoli' would bring down the terminal date of Nagda I to C. 800 B.C. He goes on to write:

> At the end of Period I the site was obviously abandoned for a while; for how long, however, is difficult to guess. But that the abandonment of the site could not have been for long is clear as is established by the continuance into the next period of certain ceramic wares, and even the microliths. This period of abandonment may be estimated at 50 years at the outside. This would place the beginning of Periof II at *circa* 750 B.C.[94]

Banerjee assigns 250 years to the deposit of Period II (about 9 layers and two structural phases) and puts it between 750 and 500 B.C.

It is, however, now clear that by the middle of the second millennium B.C. the Malwa culture was coming to its close. The precise time of this end may vary from site to site but the difference is not likely to be significant because the culture belongs more or less to a single physiographic unit. Even assuming that the Malwa culture at Nagda came to an end around 1400 B.C. and that the time-gap between this and the succeeding period is 100 years (not 50 as suggested by Banerjee), one gets the date of C. 1300 B.C. for the beginning of Nagda II.

Very little is known about Eran but there are two C–14 dates specifically mentioned as belonging to its Period IIA. There is a third date which is mentioned as belonging only to Period II but there is every possibility that it should belong to Period IIA as well. These dates are: 2990 ± 110 B.P. or 1040 + 110 B.C. (TF–326); 3220 ± 110 B.P. or 1270 ± 110 B.C. (TF–324) and 3289 ± 71 B.P. or 1339 ± 71 B.C. (P–525; this sample has been listed as coming from Period II; the particular phase has not been mentioned). When calibrated, these

[94] Banerjee, 1986: 19.

dates tend to be even earlier than the chronology suggested by us for the beginning of Period II at Nagda.[95] There are 8 other C–14 dates from Eran, all from the chalcolithic level. Six of them are earlier than the Period IIA dates and thus consistent. Two dates are found to be later and thus not consistent. But possibly it is better to depend on the evidence of six consistent dates than on two inconsistent dates.

To sum up, there is no reason at all why the beginning of the iron-bearing level in Malwa as represented at sites like Nagda and Eran should not go back to c. 1300 B.C. There is not much point in referring to the dates of Ujjain I and Besnagar I in this context because none of them possesses a chalcolithic level and thus the beginning of the iron-bearing black-and-red ware deposits of Ujjain I and Besnagar I may be later than the beginning of Period II at Nagda and of Period IIA at Eran. As regards Ujjain it is useful to note that the pre-Northern-Black-Polished-Ware phase at the site has about a 7 ft. thick deposit. Assuming the date of the Northern Black Polished Ware at the site to be c. 500 B.C., the beginning of its antecedent phase may even be as early as the beginning of Nagda II. There is nothing sacrosanct about Banerjee's assignment of 250 years to the 7 ft. thick deposit.

Iron in Early Historic Malwa

There is no lack of excavated early historic levels in Malwa but in the present context reference may be made to four sites about which some details have been published—Besnagar, Maheshwar-Navdatoli, Nagda and Tripuri. Besnagar or ancient *Vidiśā* is a predominantly early historic site and the iron objects recorded in the excavations of 1913–14 fall by and large in the closing centuries B.C. and early centuries A.D. Maheshwar-Navdatoli is famous for its chalcolithic remains but its Periods IV and V are broadly early historic and the iron objects from this level are published in full. Tripuri further east was excavated in 1952 and has a proper excavation report. The strata II–V at Tripuri in 1952 were dated between c. 300–400 B.C. and c. A.D. 400, a time-span which in generally ascribable to the objects from other sites like Maheshwar-Navdatoli and Besnagar. The early historic level at Nagda has been put between 500 and 200 B.C.

Nagda: 210 objects in all comprising the following categories: (1) sickles (long curved blades and crescentic blades), (2) hoes (socketed top for hafting and a protruding blade), (3) axes (socketed with splayed

[95] Possehl, 1988.

edge), (4) wedges, (5) bowls, (6) door plates (broad strips of iron bent to form a clamp), (7) rings, (8) washers, (9) sockets, (10) daggers and knives, (11) arrow-heads (invariably tanged), (12) spearheads (leaf-shaped with flattish cross-section, short and pointed with square cross-section and leaf-shaped with elliptical cross-section but with apparent protuberance at the base of the blade, on a solid stem), (13) a possible plumb-bob and (14) miscellaneous objects such as chisel or *khurpi*.[96]

Besnagar: (1) arrowhead, (2) sickle, (3) knife, (4) axe, (5) chisel, (6) bolt, (7) nail, (8) hook, (9) wedge, (10) staple, (11) ring.[97]

Maheshwar-Navdatoli: (1) arrowhead, (2) spearhead, (3) knife, (4) sickle, (5) hoe, (6) nail, (7) ring.[98]

Tripuri: (1) dagger, (2) knife, (3) javelin, (4) arrowhead, (5) nail, (6) clamp, (7) ring, (8) an indeterminate object but possibly a plaything.[99]

The last-mentioned type from Tripuri, of which there is only one specimen, deserves a special mention because this seems to be the only thing of its kind in iron ever found in any historical context in India. It is an iron plate about 4½ inches long with a crane, and a deer and a peacock riveted to the top. Probably there was another animal but that one is missing. The date should be about the fourth century A.D. It is worth quoting the excavator, M. G. Dikshit, in detail:

> The animals though very small are very carefully made by bending small strips of iron cut to requisite shapes and the beauty of their forms is enhanced by the simple curves with a silhouette-like effect. The crest on the top of the peacock and the dovetail effect of the body of the deer show the mastery of the smith in handling the material with a minimum of detail. Considering the age and the medium, this can certainly be called a piece of art. At the base of each rivetted animal, a small tang is left, apparently for the insertion into some wooden object to which the entire piece was fitted. This object of wood, if any, has since perished.[100]

Another object from Tripuri, a clamp from stratum IV dated 100 B.C.–A.D. 200, may be specially mentioned.

> It has two flat diamond shaped ends attached to a strong bar, circular in section, and turned to form a clamp. The ends could be fitted with nails

[96] Banerjee, 1986: 250–2.
[97] Bhandarkar, 1913–14.
[98] Sankalia, Subbarao, Deo, 1958.
[99] Dikshit, 1955.
[100] Ibid.

> to the topmost strut of the wicket in one of the door flaps, while the loop served as a latch on the other. This arrangement of clamps for closing wicket-gates is common even today and besides the present specimen, the only examples known so far seem to be from the monasteries at Sanchi.[101]

In addition to these one may cite the early historic data on iron from Ujjain (Periods II and III, covering roughly the period from c. 500 B.C. to the early centuries A.D.). Apart from 'enormous quantities of iron slag' and lumps of limonite, the objects included socketed and tanged arrowheads, spearheads, knives, blades, nails, hooks, sickles and a pair of scissors, A blacksmith's furnace was discovered in Period II.[102]

9. Vidarbha

A large number of megalithic stone-circles and habitational deposites have been excavated in the area near Nagpur, i.e. in the Vidarbha region of Maharashtra. Takalghat, Khapa, Gangapur, Mahurjhari, Naikund, Bhagimohari and Junapani are among the better known sites. The earliest level of Bhagimohari has two dates (uncalibrated): 690 ± 100 B.C. and 750 ±100 B.C.[103] The dates from the other sites fall in the same broad range. Naikund, for instance, has two dates: 620± 108 B.C. and 580 ± 103 B.C.Their calibrated ranges are 800–420 B.C. and 785–410 B.C. respectively.[104] The archaeological data suggest, on the whole, rich agricultural settlements characterized by a plethora of iron and copper objects.

The typology of early iron objects found in this region may be made clear with reference to the finds from the habitational deposits at Takalghat and the stone-circles at Khapa and Gangapur.[105] The excavator, S. B. Deo, lists about twenty types.

1. *Ladles*: 17 specimens (2 from Takalghat IB, 8 from Khapa, 7 from Gangapur)—shallow circular bowl with a straight vertical handle, the end of which is sometimes bent for a better hold.
2. *Nails*: 8 specimens (4 from Takalghat, 2 each from Khapa and Gangapur)—mostly found broken but all with convex circular heads.
3. *Dagger blades*: 4 specimens (3 from Khapa, 1 from Takalghat)—

[101] Ibid.
[102] Banerjee, 1965
[103] Possehl, 1988.
[104] Ibid.
[105] Deo, 1970: 45–50.

short tang, pointed tip and medium long blade with biconvex section—sometimes ringed base for blade.
4. *Spearhead*: 1 specimen (from Gangapur)—thin, biconvex blade, long tang with circular section.
5. *Sword*: 1 specimen (from Khapa)—'its length made it distinctive as also the sections of its long tang and blade. The tang was possibily accommodated in a wooden cover. The blade showed one edge sharp.'[106]
6. *Arrowheads*: 8 specimens (3 from Takalghat, 1 from Gangapur and 4 from Khapa)—two sub-types: tanged arrowheads with thin, biconvex, short, pointed blades and arrowheads with barbed blades.
7. *Knife*: 2 specimens (1 from Khapa, 1 from Takalghat)—slightly carved thin blade and a short tang with circular section.
8. *Chisels*: 10 specimens (6 from Gangapur, 4 from Khapa)—rectangular bar body with one end pointed and the other bevelled on both sides to achieve a thin but broad end—incipient shoulders below the pointed end—one end possibly used as a chisel and the other one used for carving.
9. *Spikes*: 2 specimens (1 each from Khapa and Gangapur)—a long thin blade tapering to a rounded point and a long rectangular tang with knobbed head.
10. *Axe with crossed fasteners*: 8 specimens (7 from Khapa and Gangapur and 1 from Takalghat)—two categories: '(a) those with elongated body with thin rectangular section, convex butt end and straight and broad working end; and (b) axes with thick broad body, convex but flat butt end, working end bevelled to a sharp convex outline and with diagonally crossing ring band fasteners'.[107]
11. *Double-edged adzes*: 26 specimens (15 from Khapa, 9 from Gangapur and 2 from Takalghat)—two categories: '(a) those with double concave outline, convex working ends and plano-convex section of the body; and (b) those with thin bodies with rectangular section, compressed middle portion and with both working ends convex'.[108] 'Deo writes: 'Keeping in view the thinness of these artifacts, it is not clear to what purpose they could have been put. The convex ends were possibly used for cutting leather, the middle portion facilitating grip because of its concavity. It is also probable that the thinnest portion was hafted into a wooden handle and the implement used at both its ends. That the cutting ends could be put to preliminary

[106] Ibid.
[107] Ibid.
[108] Ibid.

and secondary or refined uses is clear from the varying broadness of the ends.'[109]

12. *Blade with tang*: 2 specimens (1 each from Takalghat and Gangapur)—fragments.

13. *Tang pieces*: 18 specimens (7 from Khapa, 8 from Gangapur, 3 from Takalghat)—all with round sections.

14. *Bar or rod*: 2 specimens (1 each from Takalghat and Khapa)—purpose undetermined.

15. *Fish-hooks*: 2 specimens (1 each from Takalghat and Khapa)—thin rods of iron with curved up-pointed ends.

16. *Horse-bit*: 1 specimen from Khapa—two interlinked rods with circular section.

17. *Bangles*: 3 specimens from Khapa.

18. *Nailparer-cum-ear-pick*: 15 specimens (8 from Gangapur, 7 from Khapa)—'these are long thin bars of iron with circular section, one end bevelled to a sharp, thin and broad edge, the other tapering to a point, and with the body cabled or screwed for facilitating grip. The pointed end was possibly used for cleaning the ear and the broader one for nail trimming. Some specimens have a plain body.'[110]

19. *Cauldron*: 1 specimen from Khapa—flat base, rather flaring sides, thick edges and straight, circular side rings riveted with a strip at the bottom.

20. *Miscellaneous objects*: all fragments—purpose undertermined.

Some additional details can be obtained from Deo's report on Mahurjhari. An important new type is the hoe. In this report spikes have been called *śūlas*. These are of two types—those with a knobbed tang, and those with a plain tang. Deo describes them thus: 'these are long pieces with longish leaf-shaped blades and long squarish tangs which end generally in a knob. It is obvious that those with the knob cannot be inserted in a handle but could have been covered with a bamboo piece. The longest piece measures 97 cm.'[111]

Further details of horse-bits were obtained from Mahurjhari. 'Mahurjhari bridle bits are of three types: (a) those with interlocked bars having ringed ends, (b) those with interlocked bars and ringed ends having an additional double strip with smaller looped ends, and (c) similar to above but with the ringed ends having an additional single curved bar with looped ends.'[112]

[109] Ibid.
[110] Ibid.
[111] Deo, 1973: 43–54.
[112] Ibid.

Among the early historical sites excavated in the Vidarbha region Paunar deserves attention. Its Period II lies between the fourth-third centuries B.C. and second-third centuries A.D. The iron objects recovered from this period are nails (square section and round section), blades, chisels (two types: those with a broad cutting edge and those with a pointed tip), sickles, arrowheads, spearheads, rings, bangles, hooks and clamps.[113]

10. The Deccan

The first site to be considered in this context is Prakash, situated on the bank of the Tapti. The sequence here is remarkably close to that at Nagda. It has a substantial pre-Northern-Black-Polished-Ware black-and-red-ware deposit with iron beginning right from the lowest level of the deposit. There is a short hiatus of a thin gravel deposit between it and the preceding Malwa chalcolithic deposit. Towards the top there is an overlap with the Northern Black Polished Ware. The site was excavated in 1955 and there is no radiocarbon date from any of its deposits. We believe that the chronology adopted by us for Nagda II should be more or less acceptable for the iron-bearing deposit of Prakash as well. A very important point has been made by B. and F.R. Allchin in this context. They first point out that the 'Late Jorwe Chalcolithic phase continued to c. 900–800 B.C. It might be expected that the first iron objects would occur during this period. But this has not yet been shown'.[114] In the context of Inamgaon they write: 'The hollow legged pottery urns of the Iron Age are already in evidence in a burial from Inamgaon of the Later Jorwe period (c. 1000 B.C.).'[115]

The nature of iron finds at Prakash has been clearly stated by the excavator, B. K. Thapar.

> The occupation-deposit of Period II, comprising layers 44B to 33, represents an average thickness of 17½ ft. of which the upper 2 ft. or so is overlapped with the early levels of Period III. Layer 34 in the overlapped strata as also the two layers immediately preceding it, viz. 36 and 37, yielded *inter alia* sherds of the Northern Black Polished Ware. A few pottery-types, normally associated with this Ware, were, however, also found a little lower down. Of the thirty objects recovered from Period II, as many as twenty-four came from the deposits prior to the emergence of the N.B.P. Ware . . . Notable shapes from the lower deposits of Period II include: tanged arrowhead; celt-like axe-head; knife-blade; sickle; chisel-ended tanged object, possibly a carpenter's tool; lance or spearhead; and

[113] Deo and Dhavalikar, 1967.
[114] Allchin and Allchin, 1982: 327.
[115] Ibid: 342.

a ferrule, besides the ubiquitious nail. From the upper 6 to 7 ft. of deposit only a clamp and a tool of indeterminate use need mention.[116]

In the historic Periods III and IV iron objects comprised shaft-hole-axe, punch, socketed knife-blade and a bobbin-like object.

One of the most detailed lists of iron objects of the early historic in the Deccan comes from Periods IV and V at Nevasa spanning roughly the period from the second century B.C. to the early centuries A.D. Arrowheads were all tanged and with leaf-shaped blades. The long specimen of chisel was double-sloped to form a sharp cutting edge. The knife-blades were most invariably tanged pieces with blades ending in a point. The sickles fell into two groups: those with crescent-shaped curved blades and those with horizontal broad blades. The spearheads were both tanged and socketed. Adze is represented by a single piece with a flat side and a tang at an angle to the body. There is a complete piece of pick-axe with a round socket and a sharp digging point. The hooks were plain hooks with curved and sharp points. There is also a doubtful specimen of drill. Among the household utensils one may mention 'frying shovels' (flat, broad head with a long handle with rectangular section), ladles (those with a straight horizontal hold and those with a vertical hold), rimless bowls, pans or dishes, clamps (iron rods with two ends bent downwards, flat strips with riveted holds, squarish frames with downward spikes, strips with rounded edges and perforations, rods with spread heads at both ends), ringholds, nails, pegs, pounders (double-ended), lamps and possibly a seal (one face broad and biconvex but other-wise a square piece). There are also rings, door-rings, handles and unidentified objects.

The excavators of Nevasa, Sankalia, *et al.*, feel that iron was used on a large scale only in the Indo-Roman period. They point out that 'handleless ladles, frying or baking pans, frying shovels, adze and chisels have more or less exact counterparts from Taxila'. They go on to add, 'thus, the various iron objects at Nevasa are evidently the result of Roman influence'.[117] The typology of the early historic iron objects from Nevasa is more or less repeated at the other excavated early historic levels of Maharashtra. However, at Nasik (Period IIA assigned to 200–300 B.C.) one notes a new type, caltrops which 'were spread on the road with the purpose of offering obstruction to the progress of the horses and elephants of enemies'.[118]

[116] Thapar, 1964–5: 122–3.
[117] Sankalia, Deo, Ansari, Ehrhardt, 1960: 440.
[118] Sankalia and Deo, 1965.

11. South Indian Megaliths

In the south Indian protohistoric sequence iron objects first appear in the period of overlap between the neolithic and the megalithic periods. The megalithic period may be said to have ended with the beginning of the early historical period in the late centuries B.C. and the early centuries A.D. But, the practice of erecting megalithic burials lasted much longer than this, possibly even up to the mediaeval period. The megaliths by themselves, however, do not seem to display a perceptible evolutionary sequence. Specific dates have to be obtained for specific megaliths. Secondly, the excavated or otherwise recovered iron objects display an overwhelming uniformity of types, without displaying again a perceptible evolutionary and datable sequence. The south Indian megalithic iron objects thus have to be treated as a distinct assemblage which retained its homogeneity throughout its long duration.

The overlap between the neolithic and the megalithic periods (and thus the associated iron objects, mainly arrowheads and spearheads) has been found to be quite early at the site of Hallur in Karnataka. Hallur has three radiocarbon dates from this phase: TF–570: 1110 ± 108 B.C. (calibrated range 1385–1050 B.C.); TF–575: 1030 ± 103 B.C. (calibrated range 1320–1010 B.C.); TF–573: 955 ± 103 B.C. (calibrated range 1125–825 B.C.). Identical pottery has been recovered from the pit-circle graves at neighbouring Tadakanahalli, 'suggesting a similarly early date for the graves'.[119] The overlap between the neolithic and the megalithic periods has been traced also at two sites at least in Andhra—Hullikalu[120] in the Anantapur district and Pagidigutta in the Mahbubnagar district.[121] The most important breakthrough in the context of the south Indian megaliths seems to be four thermoluminiscent dates of the megalithic black, red and black-and-red pottery of a site called Kumarnahalli. These dates are 1320 B.C. (PRL–TL–46), 1380/1200 B.C. (PRL–TL–47), 1130/930 B.C. (PRL–TL–49) and 1440/1100 B.C. (PRL–TL–50).

Convenient lists of the various categories of iron objects found in the south Indian megaliths may be obtained from B. K. Gururaja Rao's *Megalithic Culture in South India* (1972) and L. Leshnik's *South Indian Megalithic Burials* (1974). However, a comprehensive

[119] Allchin and Allchin, 1982: 343.

[120] *IAR* 1978–9: 62.

[121] Ibid: 65. Neolithic-megalithic overlap has been noted also at Banahalli, Kolar district, Karnataka, *IAR* 1983–4: 42–6.

list was prepared by F. R. Allchin as early as 1954, and the following list has been basically adapted from it.[122] The general categories of what Allchin calls 'the southern irontypology' are 'general purpose and agricultural', 'stone-quarrying and working tools', 'domestic and wood-working tools', 'arms and weapons' and 'cult objects'.

1. *Flat cross banded and (?) single banded axes and (?) hoes*: there seem to be three main groups, long (12–15 inches), medium (about 8 inches) and short (about 6 inches). Allchin observes that one such axe is illustrated on a Sanchi relief and that the figures of their lengths 'compare interestingly with the large, medium large and medium small classes of stone axe'. The idea that the prototypes of these iron axes, which could also have been hafted so as to form hoes and adzes, were stone axes was first mooted by Rivett-Carnac: 'The tribes who used these weapons, having discovered the use of iron, and the place of the stone hatchet having been supplied and improved by iron, the ligatures of thong too have given way to iron.'[123]
2. *The flanged spade or (?) hoe*: two sizes, long (about 14 inches) and short (about 6 inches). Allchin treats them as ploughshares.
3. *The flanged hoe or spud*: a specialized type of spade, which also occurs among the modern Agaria as an ox-goad.
4. *The sickle and bill-hook*: these are supposed to be common forms occurring at a large number of sites.
5. *The flanged pick-axe*: rather rare but generally a robust type, with a length of about 10 inches.
6. *Stone-cutter's wedges*: functional identification was possible on the basis of modern parallels—length 6 to 8 inches.
7. *Bar wedge*: Allchin infers from the regularity of their sizes (9½–11¼ inches in length) that 'they had some special use, possibly for splitting stone or wood'.
8. *Pointed bars and crow-bars*: a number of heavy bars of 2–3 ft. in length.
9. *Chisels and adzes*: these are said to correspond closely with the finely polished small stone axes (length 4–6 inches).

10 *Knives*: small straight and curved bladed knives of oval section and tanged handles.

11. *Iron tripods*: legs riveted to the circular seat.
12. *Swords*: a wide range of iron swords with straight and leaf-shaped

[122] Gururaja Rao, 1972; Leshnik, 1974; Allchin, 1954. I am deeply indebted to Dr Allchin for making this work available to me.
[123] Allchin, 1954.

blades, tanged and sometimes with iron handles—they commonly have raised rims along the centre of the blade—one specimen from Savandurg has a tang and guard of copper.

13. *Daggers and dirks*: rather similar in shape to the knives—a specimen from the Nilgiris has a leaf-shaped blade and a bronze engraved handle.

14. *Spearheads*: four varieties have been established: (1) hollow socketed and ranging from long straight-sided tapering blades of 2 ft. 4 ins. length to curved-sided blades; (2) lances with rectangular sectioned pointed blades and hollow sockets; (3) short barbed spearheads with hollow sockets (3–6 inches long); and (4) short barbed javelins with curved sides.

15. *Arrowheads*: either barbed and tanged or socketed.

16. *Ceremonial scalloped axes*: double-angled blade attached to a back plate with holes for riveting to a shaft.

17. 'The Triśūla proper has usually a rivet to hold the side prongs. In one case at Bowenpalli a small iron model of a buffalo was attached to the shaft. The specimen from Malabar has the remains of an iron ring in a similar position, whilst one specimen from Adichanallur has a "cross bar" at the base of the triple prongs which appears to have been a similar body to the Bowenpalli specimen.'[124]

18. *Saucer hook-lamps*: a frequent type.

19. *Hooked pendants*: probably used to suspend the lamps.

20. *Objects of unknown use*: For instance, there is a reference to an 'indeterminate object with saw-like edge' at Brahmagiri.[125]

One of the major find-spots of such implements is Adichanallur where work was first carried out as early as 1876. The first definitive report on this site was published in 1902–3 and still retains its value in illustrating the detailed context of the south Indian megalithic iron tools. Adichanallur is one of the thirty-eight urn burial sites reported by Alexander Rea in the gravelly mounds adjoining the bank of the Tamraparni river in the Tinnevelly district. These sites are invariably located on waste or rocky highlands unsuitable for cultivation and there is 'almost invariably' the site of an ancient habitation in the neighbourhood of these burial mounds. A settlement site was reported by Rea about a mile north of Adichanallur, across the Tamraparni river. The lone C–14 date from Adichanallur (TF–70: 1150 ± A.D. 98) puts it in the twelfth century A.D. This may belong

[124] Ibid.
[125] Ibid.

to the terminal phase of the site. We find the following chronological estimate by Rea quite interesting:

> The Pandiyans were in possession of the Tinnevelly district from the earliest historical times. The original line of kings seems to have continued down to their conquest by Rajendra Chola in A.D. 1064. . . . To this time may be ascribed the cessation of urn burial. I do not think any of the examples are of a later date, and some of them may be much earlier. Even on the Adichanallur site, in several parts of the ground, the contents of the urns differ from those found elsewhere. Thus the bronzes are only found at certain places, skulls and complete skeletons with a few utensils at another, and so on. This may either indicate a difference in their age, or in the castes which simultaneously made use of the several parts of the burial ground.[126]

Rea's excavations covered about 5 acres of ground in the centre of the mound where large wells to contain the urns were sunk in a bed of loose quartz rock. The wells with intervening walls of rock between them varied in size from 4 ft. to 9 ft. in diameter and from 6 ft. to 12 ft. or 15 ft. in depth. In the centre of the bottom there was a small hole in which the leg of the urn was placed. The wells were filled up to the surface with gravel, leaving no trace on the surface. In the quartz bed the wells had a regular alignment but in the hard ground, outside the limits of the rocky ground, the wells were irregularly placed. Although there is an earlier report of a circle of stones being placed over the urns in some cases. Rea could not find any such indication at the sites he visited.

In addition to the burial skeletal remains (usually fractional burials but in some cases complete skeletons also) they contained, the buried urns showed pottery, iron implements, vessels and ornaments of bronze, a few gold ornaments, a few stone beads, bones and some household stone implements used for grinding curry or sandalwood. Traces of cloth and wood were sometimes found sticking to various metal objects. Husks of rice and millet were found in quite a large number of pots inside the urns. All the implements and weapons were of iron. The animals represented in bronze comprised buffalo, goat or sheep, cock, tiger, antelope and elephant. Gold diadems (some oval but others carrying thin strips extending beyond each extremity) occurred only in deeply dug and well-protected burials. Some of these diadems were plain but most of them had repoussé linear designs of dots. The repertoire of bronze objects included

[126] Rea, 1902–3: 114–40.

ornamental vase stands, bowl lids, bowls, jars, cups, sieve cups, strainers, bell-mouthed jars, necklace, ear ornaments, diadems, moulded tubes and bulbs. Some beads were made of carnelian.
Iron swords and daggers: 'All the swords and daggers have either a spike at the hilt or a curved pick-shaped piece of iron, on to which a wooden handle was attached . . . the spike is the most usual and it has been preserved from splitting the wooden handle by the use of an iron ring which exists in some examples. Those with the pick-shaped hilt have iron nails for fixing the wood, as several specimens show. The blades are of various shape. All are double-edged.'[127] The following shapes of sword blades can be discerned among the illustrated specimens: (1) concave edge and ribbed, (2) parallel-sided, angular pointed, (3) concave edges from the hilt to the middle of the length but beyond this the edges taper to form a point, (4) parallel-sided, (5) parallel sides and angular point and (6) triangular pointed. Among the daggers the common type showed spiked handle and a blade tapering to the point. In one case the hilt is concavely curved and has a flat knob at its end, pierced with a hole.
Triśūla: the illustrated specimen is 3 ft. 1½ ins. long; the handle is 2 ft. long long with a knob at the end. 'At the base of the three prongs is a cross-bar which extends out into a small knob on each side, and a curved rod underneath connects the bases of the central and side prongs. The prongs are flat and pointed at the extremities.'[128]
Lances: less than 4 ft. in each case, with a cross-piece below the hilt.
Spears: 1–3 ft. long, tapering blade, spiked handle.
Arrows: all barbed, all socketed.
Javelins: maximum length 6¾ ins.
Sacrificial daggers: maximum length about 6 ins.; a stiletto-shaped implement with spear point. 'A similar implement is now a days used to impale sacrificial fowls.'
Hatchets or axes: thick plate of flat metal, flat rounded cutting edge, flat oval diagonally attached detached rings, maximum length more than 8 ft.
Mammutis or spades: numerous, made of thick metal, the butt end bent inwards to form a socket for handle, square, rounded or convex digging ends, sides curved or straight, maximum length more than 1 ft.
Large iron hangers: length 1–2 ft. 'They have a strong broad suspending ring at the top of a vertical rod of thick square section,

[127] Ibid.
[128] Ibid.

which again has 4 large hooks of various designs at the bottom. Close to the top suspending ring, a series of from four to eight arms or ribs branch around, outwards and downwards, resembling in this respect the ribs of an ordinary umbrella. These also are terminated by hooks of thin flat metal at the extremities.'[129] *Saucer-lamps*: (4 ins. in diameter, height 6¾ ins.) circular arched bar with a hole for suspender at the top of the arch.

Iron beam rods: (1 ft. 7 ins. long, diameter of about 1 in.) long rods of round metal with an elongated oval bulging in the middle and a knob at each end.

Chisels: (maximum length over 6 ins.) cutting edge sharp.

Tripods: formed of a ring of flat metal resting on three legs which partly curve outwards and have a bent rest at the foot.

Miscellaneous: cylindrical handle-ring, reaping hooks, numbers of others of various kinds in fragmentary condition.

Adichanallur is an exceptional site. The objects from the excavated pit-circles at Brahmagiri may be more representative of the south Indian megalithic iron finds. The pit-circles at Brahmagiri yielded 7 tanged knives or daggers, a barbed arrowhead, a thin dish-like object, a fragmentary ring with 2 nails affixed to it, a chopper-like object with a long handle and 3 spears, besides 20 fragmentary and highly decayed objects.[130]

The iron objects found in Periods II and III at Maski may be considered representative iron finds at a south Indian habitational site. Period II is megalithic, dated from c. 200 B.C. to the middle of the first century A.D. with a reasonable margin of a century on the earlier side. Period III is put between the middle of the first century A.D. and the third century A.D. The total number of iron objects in Period II is thirty-two, out of which ten are from the megalithic burials. From the habitational deposit came the types like chisel, tanged arrowhead, bangle, ferrule, sickle, nail, etc. Period III yielded nails, an object with long tang and flattened lower end, knives or daggers and a narrow vase.[131]

12. Orissa

Sisupalgarh is the primary site in this context. Period I at the site did

[129] Ibid.

[130] Wheeler, 1947–8.

[131] Thapar, 1957. In Andhra the early historical level of Satanikota, dated c. first century A.D. has yielded a large number of iron objects including daggers, spearheads, nailparers, clamps, chisels, etc. See Ghosh, 1986.

not yield any iron object but this may be due to the fact that a very small area was excavated in the lowest levels of the black-and-red ware deposit. Period IIA (200 B.C.–A.D. 100) showed nails, clamps and rings. Period IIB (100–A.D. 200) objects included nails, chisels, staples, borers, tanged spearheads, bangles, knife-blades, ferrules, chopper-like objects, etc. The objects from Period III comprised the following: nails (knob-head, flathook head, flat circular head), chisels, fish-hooks, bangles, latches with provision for a central nail, staples, rings, shallow dishes, barbed arrowhead, four-edged and tanged arrowhead, three-edged tanged arrowheads, elongated, tanged, four-edged arrowheads, leaf-shaped arrowheads with flattened section, socketed spearheads, awls, harpoons of oblong section, lances, hoes, sickles, tripod stands, caltrops, etc. The excavator, B. B. Lal, pointed out that caltrops occurring at the Roman military sites of the early centuries A.D. were a notable find at Sisupalgarh.[132]

II. TECHNICAL STUDIES OF ANCIENT IRON OBJECTS

In 1926 J. Newton Friend provided an excellent summary of most of the evidence available till that date.[133] Mainly he took up the results of analysis of the Delhi Pillar (fourth century A.D.), the Dhar Pillar (Vincent Smith puts it in the fifth century A.D. but Newton Friend cites this date as A.D. 1304) and the iron beams of the famous temple of Konarak (thirteenth century A.D.) For the Delhi Pillar he cited the results of analysis undertaken by Robert Hadfield:[134] carbon–0·080 per cent, silicon–0·046 per cent, sulphur–0·006, phosphorus–0·114 percent, manganese–nil, nitrogen–0·032 per cent, total of elements other than iron, including copper–0·034 per cent, iron–99·720 per cent, total–100 ·032 per cent, density–7·81, Brinnel Hardness–188.

> It will be noticed that the material is an excellent type of wrought iron, the sulphar being particularly low (0·006 per cent), indicating that the fuel used in its manufacture and treatment must have been very pure (probably charcoal). The phosphorus is 0·114 per cent. There is no manganese present . . . The iron was ascertained by actual analysis, and not 'by difference'.[135]

A major aspect of the Delhi pillar is its general freedom from corrosion. Although the high phosphorus and low carbon, sulphur

[132] B. B. Lal, 1949.
[133] Friend, 1926: 142–55.
[134] Hadfield, 1912, also Hadfield, 1925.
[135] Hadfield, 1912.

and manganese content all tend towards reduction of corrodibility, they are not sufficient to explain the general immunity of the pillar from corrosion. This resistance to corrosion may also be due to the highly polished surface condition of the pillar which, according to an ancient custom, could be anointed with butter on certain religious occasions. On its structure Newton Friend writes the following:

> The structure consists of large grains of ferrite with a very small portion of cementite, sometimes located in the grain junctions and occasionally in the ground mass. A smaller grain structure independent of the larger one is more or less faintly traceable. There are also a large number of small lines, which at higher magnification are shown to have a regular formation and appear related to the smaller grain structure, and may be due to secular effects—that is ageing.
>
> On reheating a portion of the specimen to 900°C, the whole of this fragmentary grain structure is made to disappear leaving only large clear etching ferrite grains.
>
> The specimen is remarkably free from slag and other inclusions, and is evidently a very pure iron.[136]

Discussing the general freedom of Indian iron from rust, Newton Friend cites the opinion of one Mr Wallace from Bombay: 'I have seen native-made iron forged on a stone anvil, and have observed that it does not rust like English iron when exposed to the weather. The ironwork of the car on which the gods of the Kulu valley take the air has a fine brown patina and no rust flakes. It is all charcoal iron.'[137]

Some insight into the technical aspect of the Delhi Pillar was provided by Andrew McWilliam in 1920 in the context of the pre-industrial iron manufacture studied by him in the village of Mirjati near Jamshedpur.

> The analyses of the Mirjati iron and of the Delhi Pillar are very much alike. In the Mirjati iron the silicon varied in different lots of drillings from 0·09 to 0·22 per cent, no doubt depending largely on the amount of cinder that happened to be included. The sulphur in different samples varied from a mere trace to 0.·036 and the phosphorus from 0·14 to 0·20. In agreement with the indications given by the microscope the carbon was variable, a second lot of drillings, for example, giving 0·2 per cent carbon.
>
> The manganese in all the samples showed as nil or a mere trace. The ore contained 0·9 per cent Mno and the cinder 1·9 per cent Mno, so it is evident that manganese is not reduced under the conditions of the process, and as

136 Friend, 1926: 143–6.
137 Ibid.

> by other processes for the manufacture of wrought iron there is generally some manganese present the absence of manganese is nearly characteristic of iron made by this ancient process and gives confirmatory evidence in support of the view that the iron for the Delhi Pillar was made by a process essentially the same.[138]

McWilliam further wrote:

> The author has twice carefully examined this pillar, and although whilst approaching it the pillar looks quite smooth, on a close inspection by anyone accustomed to examine steel bars for rokes and similar flaws, the weld marks on the Delhi Pillar are quite clear and unmistakable. The similarity between the compositions of the Delhi Pillar and the Mirjati and similar irons, especially the almost complete absence of manganese, points strongly to the material having been made by the same old direct process still in ordinary use. That the pillar is made of pieces of wrought iron welded together was settled many years ago. . .[139]

In more recent years there have been at least two studies of the corrosion-resistance aspect of the Delhi Pillar, reaching more of less the same conclusion as above.[140]

The composition of the Dhar Pillar was the following: carbon–0·02 per cent, phosphorus–0·28 per cent, iron–99·60 per cent. Apart from iron, the Konarak iron beams contained the following: carbon–0·110 per cent, silicon–0·100 per cent, sulphar–0·024 per cent, phosphorus–0·015 per cent, manganese-trace.

> . . . portions of the specimen more distant from the cracks showed a fairly uniform structure, typical of mild steel containing rather less than 0·15 per cent carbon; whilst portions bordering on the holes and cracks were free from pearlite. These observations lend support to the statement by Graves that the method of manufacture consisted in welding up small blooms, and it would appear that decarburisation took place during the welding. The metal was found to be very soft, Brinnel hardness no. 72. Using the magnifier hammer the hardness number varied from 31 to 28 in different parts of the specimen.[141]

The Heliodorus Pillar (made of stone) was erected at the famous early historic site of Besnagar or ancient Vidiśā about 140 B.C. The pillar rested on a stone block. To make the pillar quite perpendicular, two pieces of iron were used and two stone chips had been driven

[138] McWilliam, 1920: 163–4.

[139] Ibid: 165.

[140] Bardgett and Stanners, 1963; Lahiri, Banerjee and Nijhawan, 1963.

[141] For Dhar Smith, 1897; also Friend, 1926: 148–50. For Konarak beams, Graves, 1912; Friend and Thornycroft, 1924. The quotation is from Friend, 1926: 153.

in between them. In his report on the work at Besnagar, D.R. Bhandarkar wrote that at the suggestion of Sir John Marshall he sent these two iron pieces for examination to Sir Robert Hadfield. Bhandarkar cites Hadfield's report on these specimens:

> I have now examined the specimen found at the base of the . . . column. This was received in such an oxidised condition that it was almost impossible to get a proper analysis. The composition was found to be as follows:-

C	Si	S	P	Mn	Cr	Ni	Fe
·70	·04	·008	·02	·02	trace	trace	99·5

> I also made physical tests, of which the following are the particular: There are some parts of the specimen from which one could obtain fracture, which showed fine crystalline to fibrous, rather brittle. The Brinnel ball hardness number was 146 after all the scale had been removed. On cutting the specimen through with a saw there was found to be a fair proportion of the original metal still unoxidised. The metal is folded over with scale in between the fold.

According to Bhandarkar, two months later Hadfield stated the following: 'I have now made a further examination of this specimen and somewhat to my astonishment find that it contains ·7 per cent carbon, which means of course that the material is steel and could be hardened by heating and quenching in water. This is the first specimen I have found of really ancient date in which there has been found a considerable percentage of carbon. The specimen therefore becomes of unusual interest.'[142]

An extensive technological examination of excavated iron objects was undertaken by John Marshall at Taxila. We give the results detailed by Hadfield below (we omit the antiquity numbers of the specimens but do not change their serial numbers).[143]

1. Double-edged sword—15 ins. length—average Brinnel hardness 235—high carbon material; about 1·5–1·7 per cent carbon.
2. Fragmentary sword with copper guard—9·9 ins.—average Brinnel hardness 238—high carbon material; about 1·5–1·7 per cent carbon.
3. Dagger, end of blade near haft, micro-piece from point end of dagger—9·4 ins.—average Brinnel hardness 164—medium or low carbon material severely decarburised on the surface.
4. Dagger—9·6 ins.—average Brinnel hardness 131—probably iron.
5. Adze for carpenters—5·75 ins.—away from edge—average Brinnel hardness 240—high carbon steel (1·23 per cent).

142 Bhandarkar, 1913–14.

143 Hadfield, 1951: 536–7, 562–3.

6. Axe, near cutting edge—5·75 ins.—average Brinnel hardness 125—iron.
7. Chisel—5·25 ins.—average Brinnel hardness 107—iron.
8. Knife—5 ins.—average Brinnel hardness 120—iron.
9. Three-flanged arrowhead—4·1 ins.—average Brinnel hardness 182—medium or low carbon steel.
10. Double-edged spearhead—7·5 ins.—average Brinnel hardess 113—iron.

Specimen numbers 5 and 6 in the above list were subjected to chemical analysis which is given as follows.

	C	Si	S	P	Mn
(5)	1·23	0·28	0·004	0·024	0·01
(6)	0·10	0·03	0·004	0·062	nil

Micro-examination was done of three specimens.
1. Structure consists of small slightly elongated grains of ferrite and spheroidal carbide, the result of decomposition of the pearlite. The grains are outlined by cementite. Traces only of decarburisation round the outer surface. Non-metallic inclusions fairly small and comparatively few.
2. Structure similar to specimen no. 1, but the cell-walls of cementite thicker and grains larger and more elongated. Slight partial decarburisation of the surfaces. Non-metallic inclusions similar to those in specimen no. 1.
3. Shows a core of small ferrite and pearlite grains, corresponding to the material of 0·15–0·25 per cent C surrounded by a skin of 0·1–0·25 mm in thickness of coarse columnar-shaped grains of ferrite. Non-metallic inclusions moderately large and numerous.

It is interesting to note that all the five specimens with high or low to medium carbon and which can be called steel all come from Sirkap stratum II, i.e., the Saka-Parthian period of roughly the first century A.D.

Marshall also sent to Hadfield's laboratory in Sheffield six ingots which were spindle-shaped and roughly hexagonal in section, weighing individually from 1 lb 11½ oz. to 3 lb 7¾ oz. Analyses and tests were done on one of the ingots (6 ins. long and 1·9 ins. at its greatest width) of medium weight (2 lb. 15 oz). The results of the chemical analysis were the following:

C	Si	S	P	Mn	Fe
0·10	0·03	0·019	0·077	Trace	99·6

A complete longitudinal section showed the ingot to be very porous, some of the holes being ½ in. in maximum dimension. Various hardness, tensile and shock tests were carried out. The specimen was considered 'roaky' and the standard diamond pyramid hardness figures over the section varied between 76 and 129, with an average of 95.

Hadfield's general remarks were as follows:

> (1) In their chemical analysis these ingots were fairly representative of the many Indian iron specimens, obtained from various locations, which we have examined.
> (2) Phosphorus contents as low as 0·015 per cent were found in chippings from the iron beams at Konarak, and as high as 0·28 per cent in the iron pillar or beam at Dhar.
> (3) Sulphur analysed as low as 0·002 per cent in tools found in Dekhan, but as high as 0·024 per cent in the Konarak beams, with a general average of 0·009 per cent.
> (4) The sulphur, 0·019 per cent in the present ingots, therefore, is rather above average.
> (5) It may be recalled, however, that the iron implements from Taxila examined on a previous occasion contained from 0·024 to 0·064 per cent of phosphorus, and only 0·004–0·005 per cent of sulphur.
> (6) Apparently, therefore, the ingot now examined is individually, as regards its sulphur content, rather higher than the general average of the iron produced at Taxila.
> (7) Our examination does not indicate any special qualities in this iron above that produced in other parts of India, recognizing, however, that judged by ordinary standards, Indian iron is in general of excellent purity.[144]

In their monograph entitled, *Coinage in Ancient India, a Numismatic, Archaeochemical and Metallurgical Study of Ancient Indian Coins*, (1968) Satya Prakash and Rajendra Singh offered the details of a chemical and metallurgical study undertaken by them on six iron objects from Kauśāmbī. The list of the objects with their approximate dates is given below.

1. Arrowhead—100 B.C. 2. Arrowhead—300–200 B.C. 3. Arrowhead—100 B.C.–A.D. 500 4. Arrowhead—A.D. 480 5. Arrowhead—rectangular cross-section—395–325 B.C. 6. Iron piece—c. 200 B.C.?

Table 2 on P. 92 gives the results of chemical analysis.

The results of their metallographic examination are as follows:

> The microstructure of sample no. 1 shows the structure of a low carbon steel having polyhedral grains of ferrite with pearlite at the grain boundaries. Lot of inclusions of Feo, Si02 are present.

[144] Ibid.

Table 2
Chemical Analysis of Iron Objects from Kauśāmbī

	Total iron	*Si*	*Ni*	*P*	*S*	*C*
1.	89·23	0·06	–	0·04	0·03	0·08
2.	72·13	0·03	–	0·23	0·03	0·11
3.	90·17	0·05	tr.	0·21	tr.	0·74
4.	67·29	0·18	0·01	0·30	0·05	0·65
5.	63·12	0·20	–	0·09	0·01	0·41
6.	60·08	0·15	–	0·12	tr.	0·12

> The microstructure of sample no. 2 shows a very mixed deformed structure showing flowed grains filled with brown matrix, and a lot of pitting is there due to corrosion.
> Sample no. 3 has shown the structure of a high carbon steel. The structure shows the grains of pearlite and ferrite present at the grain boundaries.
> The remaining samples were highly corroded, and showed structures similar to that of sample no. 2.[145]

H. C. Bharadwaj studied six specimens (all dated approximately 600–400 B.C.) from Rajghat: (1) blade, (2) arrowhead, (3) broken nail, (4) piece of a rod, (5) fragmentary piece and (6) bent nail. He undertook both chemical and spectographic analyses of these specimens. He arrived at the following conclusions:

> From the chemical analysis it looks that silica, alumina, lime and magnesia are invariably present as impurities. The percentage of silica ranges from 0·42 to 3·8 and that of alumina, from 0·15 to 2·01. The percentage of lime is in the range of trace to 1·2 and magnesia between 0·08 to 0·80.
> In addition to these the oxides of phosphorus are also present in the range of 0·15 to 0·22 per cent. Percentage of sulphur is found to vary within 0·03 to 0·19.
> Except for specimen no. 6, the carbon content of the specimens lies between 0·12 to 0·42 per cent. Specimen no. 6 contains 1·4 per cent carbon which is much higher than others.
> From the figures mentioned above we can say that specimens 1 to 5 were malleable and could be welded. They would not harden greatly when cooled. However, these objects might have been naturally carburized during forging, when these were placed in charcoal fire. Specimen no. 6 contains higher amount of carbon and only little amount of slag and is different from the others. These specimens could have been made by forging the spongy iron produced by direct reduction of the oxide ores like haematite or magnetite.[146]

[145] Prakash and Singh, 1968: 529–32.
[146] Bharadwaj, 1973.

Bharadwaj also infers that no flux was used during smelting. He examined four of these specimens metallographically.

> (No.2) The microstructure shows ferrite and slag mixture. The metal has been worked. There is a banded structure due to forging. A large number of glossy and oxide slags are visible. There is a central cavity due to gassy evolution during solidification. There are some cavities due to dislodging of slag particles.
> (No.3) The microstructure reveals banded structure. There are alternate metal and slag rich areas due to hot working. Some glassy inclusions are also seen.
> (No.4) The microstructure shows slag inclusions, cavities and black and white lamellar structure.
> (No.6) The microstructure shows equiaxed ferrite grains, but there is no evidence of pearlite. Grains are of uniform size. There is evidence of small amount of iron carbide at the grain boundaries. This sample shows purer iron with very few slag particles.[147]

P. K. Chattopadhyay analysed a small iron knife dated c. 200 B.C. from Kangra in Himachal Pradesh. The structure contained martensite and thus the specimen was interpreted as containing the evidence of quenching.[148] Four rod-like iron objects from the middle and upper phases of the Painted Grey Ware at Atranjikhera were scientifically studied by O. P. Agrawal. He also analysed five slag samples from these deposits. The analysed iron objects were given the following sample numbers—sample 1: upper phase; sample 2: middle phase; sample 3: middle phase; sample 4: middle phase. Table 3 on P. 94 gives results of their chemical and spectographic analyses.

From the metallurgical examination it appeared that 'these objects were first made of wrought iron and later on carburized by some technique whereby the surface or the zone nearest to the surface turned into carbide. This happens when the iron object is kept on charcoal bed for long time at high temperature'.[149]

The sickle-like object found at Barudih in eastern India (Singhbhum in Bihar) was examined by P. K. Chattopadhyay. It was found to be low carbon steel, thoroughly forged and cooled slowly in the air.[150] In V. C. Athavale's analysis of iron objects from Prakash, a shaft-hole axe dated around 200 B.C. showed 'equiaxed ferrite grains and a

[147] Ibid.
[148] Chattopadhyay, 1984.
[149] O. P. Agrawal, 1983.
[150] Information from Shri P. K. Chattopadhyay.

Table 3
Chemical and Spectographic Analyses of Iron Objects at Atranjikhera

	Sample 1	*Sample 2*	*Sample 3*	*Sample 4*
Fe_2O_3	89·36	—	—	—
SiO_2	0·53	1·80	1·51	1·32
Al_2O_3	1·33	·82	·96	1·25
CaO	0·60	1·31	·82	·73
MgO	0·12	0·19	·71	·62
Cu	—	—	—	—
Ni	—	—	—	—
TiO_2	Tr.	Tr.	—	Tr.
P_2O_5	0·33	0·21	0·18	0·21
S	0·08	0·10	0·09	0·12
C	0·33	0·18	0·26	0·31
MnO_2	Tr.	Tr.	Tr.	Tr.
Co	—	—	—	—
Zn	—	—	—	—
As	—	—	—	—

small amount of pearlite at the grain-boundaries'.[151] Athavale studied seven specimens, only two of which were found to have metallic cores. Chemical analysis suggests the general use of wrought iron. An iron axe from Mahurjhari showed 0·9 per cent carbon.[152] Regarding an iron spear from Takalghat-Khapa, which is found to possess 99·6 per cent iron, it is stated that 'it is very hard and may be a variety of steel'.[153]

A trident from Raigir, a megalithic site in Andhra was reported on 'by Hadfield, who notes a carbon content of ·07 and ·85 per cent. The structure of the specimen was not uniform and the content rose in one place to 1·3 per cent and fell in another to ·1 per cent'.[154] Tadakanahalli in Karnataka is a megalithic site near Hallur and possibly equally early. O. P. Agrawal studied the material from this site and found both ferrite and pearlite.[155] In the context of Naikund, a megalithic site in Vidarbha, V. D. Gogte's finding is that the iron pieces of plain medium carbon steel were forged in high temperature.[156]

[151] Athavale, 1964–5.
[152] Joshi, 1973.
[153] Munshi and Sarin, 1970.
[154] Allchin, 1954.
[155] Information from O. P. Agrawal.
[156] Deo, 1982.

K. T. M. Hegde studied an early historic iron-smelting site called Dhatwa in the Tapti valley, Gujarat. He studied, among other things, the manufacturing process of a hoe. He concluded that it was done in two stages: 'First, the red hot bloom was forged into thin sheets on an anvil near an open hearth. In this process the surface of the sheets was carburized and casehardened. Secondly, a number of these sheets were joined laterally, one by one, by forge-welding. Finally, the whole mass was further forged to shape it into a hoe. The laminated structure in the metal indicates the lines of lateral welding.'[157]

Another important aspect of Hegde's study was the fact that, by undertaking analytical study of the ore pieces found at the site, he could infer that 'haematite was produced by roasting limonite at the site'. He further stated: 'The evidence of roasting of the ore is not without significance. In roasting limonite loses much of its water content together with carbon dioxide and other volatile components in the ore. Therefore iron oxide content of the ore increases weight per weight and it becomes more porous, fragile, easily crushable and suitable for reduction.'[158]

N. C. Ghosh, A. K. Nag and P. K. Chattopadhyay (1989) have published the technical details of an iron dagger found in the period II of a chalcolithic site called Hatigra in Birbhum, West Bengal. According to Chattopadhyay's study, the metal used is manganese-free and the carburisation of the surface was done in prolonged heating, around 1200°C. (See 'The Archaeological Background and Iron Sample from Hatigra, *Puratattva* 18: 21–7).

B. Prakash (*Puratattva*, 1989–90: 118–22) discusses the forging techniques of the Dhar pillar. Lothal Report 2: 657 shows a lump with 39·1 per cent iron and a fragment with 66·1 per cent iron and 9·3 per cent copper. Dr. B. B. Lal writes about this Harappan 'fragment': 'It would thus be seen that the specimen No. 15112 is made of iron containing a small proportion of copper namely 9·3 per cent'. He adds, 'There is no doubt that the use of gold, silver, copper, *iron* (italics ours) and bronze was fully understood'. S. R. Rao and all of us overlooked this find, and Nayanjot Lahiri who drew my attention to this gets all credit as Sahi gets the credit in the context of Ahar.

[157] Hegde, 1973
[158] Ibid

CHAPTER FOUR

The Literary Sources

I. DATA

The Ṛgveda (RV)

The keyword in this context is *ayas* which occurs in the following places in the *RV*:

1. 1.52.8: *vajramāyasam*—(Indra's) thunderbolt of *ayas*.[1]
2. 1.56.3: *āyasaḥ*—(Indra) clothed in armour made of *ayas*.[2]
3. 1.57.3: *haritaḥ na ayase*—(Indra's) thunderbolt of *ayas*.[3]
4. 1.58.8: *āyasībhiḥ*—'forts of *ayas*' in prayer to Agni.[4]
5. 1.80.12: *vajraḥ āyasaḥ*—(Indra's) thunderbolt of *ayas*.[5]
6. 1.81.4: *vajramāyasam*—(Indra's) thunderbolt of *ayas*.[6]
7. 1.88.5: *ayodaṃṣṭrān*—(Marut) armed with *ayas* weapons.[7]
8. 1.116.15: *jaṃghāmāyasīm*—(Indra) gave her (Viśpalā who lost her leg) a leg of *ayas*.[8]
9. 1.121.9: *āyasam*—(Indra's) thunderbolt of *ayas*.[9]
10. 1.163.9: *ayaḥ asya pādā*—his feet are of *ayas* (context of sacrificial horse).[10]

[1] Wilson, 1850–88, vol. I: 143; Tilak Maharashtra University (hereafter cited as TMU), 1933–, I: 368.

[2] Wilson, 1850–88, I: 153, TMU, 1933–, I: 393.

[3] TMU, 1933–, I: 397.

[4] Griffith, 1896–7, I: 80; TMU, 1933–, I: 403.

[5] Wilson, 1850–88, I: 206; TMU, 1933–, I: 502.

[6] Wilson, 1850–88, I: 206; TMU, 1933–, I: 507.

[7] Wilson, 1850–88, I: 226; TMU, 1933–, I: 585.

[8] Wilson, 1850–88, I: 311; TMU, 1933–, I: 721.

[9] Wilson, 1850–88, I: 328; TMU, 1933–, I: 762.

[10] Wilson, 1850–88, II: 123; TMU, 1933–, I: 974.

11. 2.20.8: *āyasīḥ*—cities of *ayas* (of the *dasyus*).[11]
12. 4.2.17: *ayaḥ . . . dhamaṃtaḥ (yathā karmārā ayo bhastreṇa dhamanti tadvat*—Sāyaṇa's commentary)—performers of good works . . . have freed their birth from impurity, as (a smith heats) *ayas*.[12]
13. 4.27.1:*āyasīḥ*—cities of *ayas* (of the *dasyus*).[13]
14. 4.37.4: *ayahsiprah*—jaws of *ayas* (context of horse).[14]
15. 5.30.15: *ayasmaya gharma*—vessel of *ayas*.[15]
16. 5.62.7: *ayo asya sthūṇā*—its columns are of *ayas* (chariot of Mitra-Varuṇa).[16]
17. 5.62.8: *ayah sthūṇaṃ*—*ayas*-pillared (chariot of Mitra-Varuṇa).[17]
18. 6.3.5: *tejaḥ ayasaḥ*—(Agni's) splendour like the edge of *ayas*.[18]
19. 6.47.10: *ayasaḥ na dhārāṃ*—(Indra) sharpen my thought as 'twere a blade of *ayas*.[19]
20. 6.71.4: *ayaḥ hanūḥ (hiranmaya hanū*—Sāyaṇa)—context *Indrasomou*.[20]
21. 6.75.15: *ayomukhaṃ*—(shaft) with *ayas* mouth.[21]
22. 7.3.7: *sataṃ pūrbhirāyasībhih*—Agni–guard us . . . with those boundless glories as with a hundred fortresses of iron.[22]
23. 7.15.14: *āyasī*—Agni–be thou a mighty iron fort to us.[23]
24. 7.95.1: *āyasī*—the stream Sarasvati with fostering current comes forth, our sure defence, our fort of iron.[24]
25. 8.101.3: *ayaḥ śirṣā*—*ayas*-topped.[25]
26. 9.1.2: *ayohatam*—(Indra) his place, his *ayas*-fashioned home.[26]
27. 9.80.2: *ayohatam*—(Soma) his *ayas*-fashioned home.[27]

[11] Wilson, 1850–88, II: 258; TMU, 1933–, II: 83.
[12] Wilson, 1850–88, III: 121; TMU, 1933–, II: 510.
[13] Max Muller, 1965, I: 286.
[14] Griffith, 1896–7, I: 443; TMU, 1933–, II: 654.
[15] Max Muller, 1965, I: 326.
[16] Griffith, 1896–7, I: 533; TMU, 1933–, II: 535.
[17] Griffith, 1896–7, I: 534; TMU, 1933–, II: 934–5.
[18] Griffith, 1896–7, I: 558; TMU, 1933–, III: 12.
[19] Griffith, 1896–7, I: 610; TMU, 1933–, III: 161.
[20] Max Muller, 1965, I: 427.
[21] Griffith, 1896–7, I: 647; TMU, 1933–, III: 261.
[22] Griffith, 1896–7, II: 5; TMU, 1933–, III: 276.
[23] Griffith, 1896–7, II: 15; TMU, 1933–, III: 299.
[24] Griffith, 1896–7, II: 90; TMU, 1933–, III: 488.
[25] Max Muller, 1965, II: 181.
[26] Griffith, 1896–7, II: 269; TMU, 1933–, IV: 1.
[27] Griffith, 1896–7, II: 337; TMU, 1933–, IV: 140.

28. 10.53.9: *paraśuṃ svāyasaṃ*—axe of *ayas*.[28]
29. 10.87.2: *ayah daṃṣtraḥ*—(Agni) with the teeth of *ayas*.[29]
30. 10.96.3–4: *āyasaḥ*—(Indra) his is that thunderbolt of iron, golden hued, gold-coloured, very dear, yellow in his arms.[30]
31. 10.96.8: *āyasaḥ (ayomaya hṛdayam)*—at the swift draught the Soma drinker waxed in might, the Iron one with yellow beard and yellow hair.[31]
32. 10.99.6: *ayah agrāya*—struck down the boar with shaft whose point was iron.[32]
33. 10.99.8: *ayaḥ apāṣṭiḥ*—beak of *ayas*.[33]
34. 10.113.5: *vajramāyasaṃ*—thunderbolt of *ayas*.[34]

We shall give our own observation on the meaning of *ayas* in the *RV* in the subsequent section on 'discussion'; here we shall only note how scholars have interpreted this term. Vaman Shivaram Apte in *The Practical Sanskrit-English Dictionary* explains the term as 'iron, steel, gold, a metal in general and aloe wood'.[35] In *A Comparative Dictionary of the Indo-Aryan Languages* by R. L. Turner it is listed as 'metal, iron'.[36] The same listing is done in Carl Cappeller's *A Sanskrit-English Dictionary*[37] and Arthur A. Macdonell's *A Sanskrit-English Dictionary*.[38] Theodore Benfey in his *A Sanskrit-English Dictionary* puts *ayas* as iron.[39] The same meaning is attached by T. Goldstucker in his *A Dictionary of Sanskrit and English*[40] and by Otto Bohtlingk and Rudolph Roth in their *Sanskrit-Worterbuch*.[41] Monier Williams in his *A Sanskrit-English Dictionary* puts it as 'iron, steel, gold, metal'.[42]

A number of references which refer to miscellaneous metal objects and metallurgical activities have been interpreted by M. N. Banerjee

[28] Max Muller, 1965, II: 309.
[29] Griffith, 1896–7, II: 510; TMU, 1933–, IV: 593.
[30] Griffith, 1896–7, II: 531; TMU, 1933–, IV: 648.
[31] Griffith, 1896–7, II: 532; TMU, 1933–, IV: 649.
[32] Griffith, 1896–7, II: 537; TMU, 1933–, IV: 663.
[33] Max Muller, 1965, II: 362.
[34] Ibid: 374.
[35] Apte, 1957, I: 210–11.
[36] Turner, 1966: 26.
[37] Cappeller, 1891: 39.
[38] Macdonell, 1893: 26.
[39] Benfey, 1866: 46.
[40] Goldstucker, 1856: 433.
[41] Bohtlingk and Roth, 1855.
[42] Williams, 1872: 79.

Illustrations

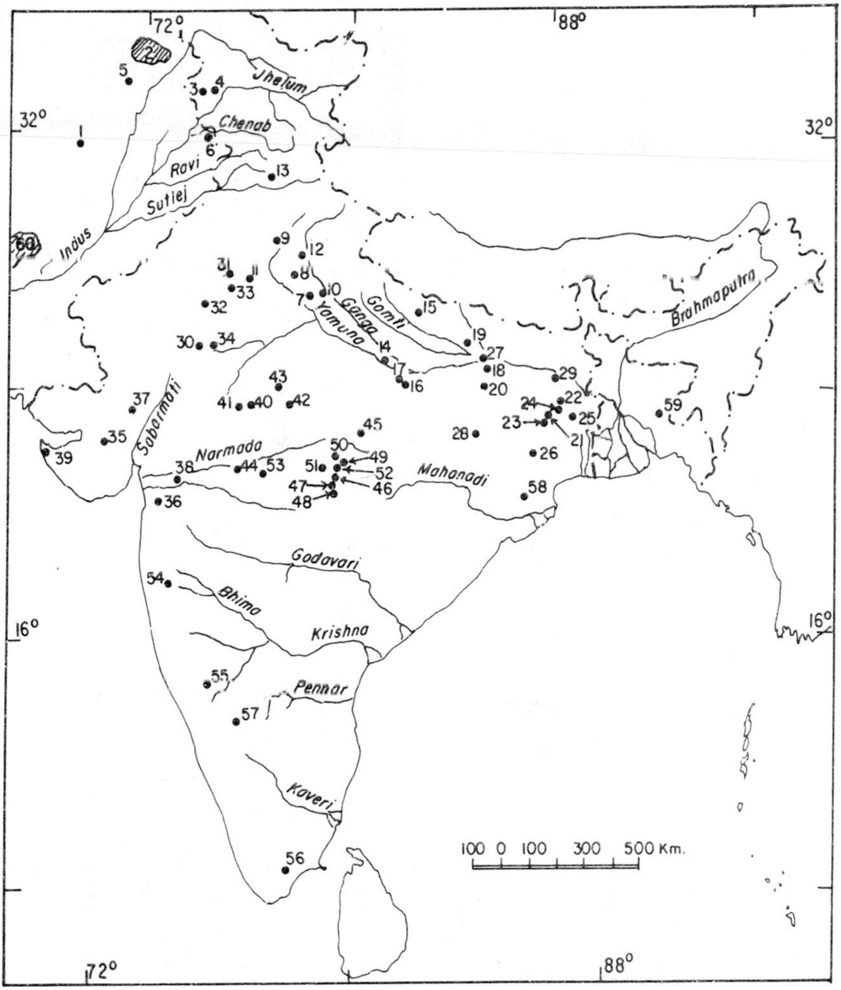

Fig. 1. Map Showing Major Sites Mentioned in the Text.

1. Pirak, 2. Gandhara Graves, 3. Saraikhola, 4. Taxila, 5. Shaikhan Dheri, 6. Tulamba, 7. Atranjikhera, 8. Alamgirpur, 9. Hulas, 10. Jakhera, 11. Noh, 12. Hastinapur, 13. Rupar, 14. Kausambi, 15. Ganwaria, 16. Koldihawa, 17. Panchoh, 18. Sonpur, 19. Chirand, 20. Taradih, 21. Pandu Rajar Dhibi, 22. Mahisdal, 23. Bharatpur, 24. Bahiri, 25. Hatigra, 26. Barudih, 27. Pataliputra, 28. Saradkel, 29. Karkhup, 30. Ahar, 31. Bairat, 32. Sambhar, 33. Rairh, 34. Nagari, 35. Nagara, 36. Dhatwa, 37. Devnimori, 38. Timbarva, 39. Dwaraka, 40. Nagda, 41. Ujjain, 42. Besnagar, 43. Eran, 44. Navdatoli, 45. Tripuri, 46. Takalghat, 47. Khapa, 48. Gangapur, 49. Mahurjhari, 50. Naikund, 51. Bhagimohari, 52. Junapani, 53. Prakash, 54. Nevasa, 55. Hallur, 56. Adichanallur, 57. Brahmagiri, 58. Sisupalgarh, 59. Wari-Bateshwar, 60. Baluchistan cairn burials

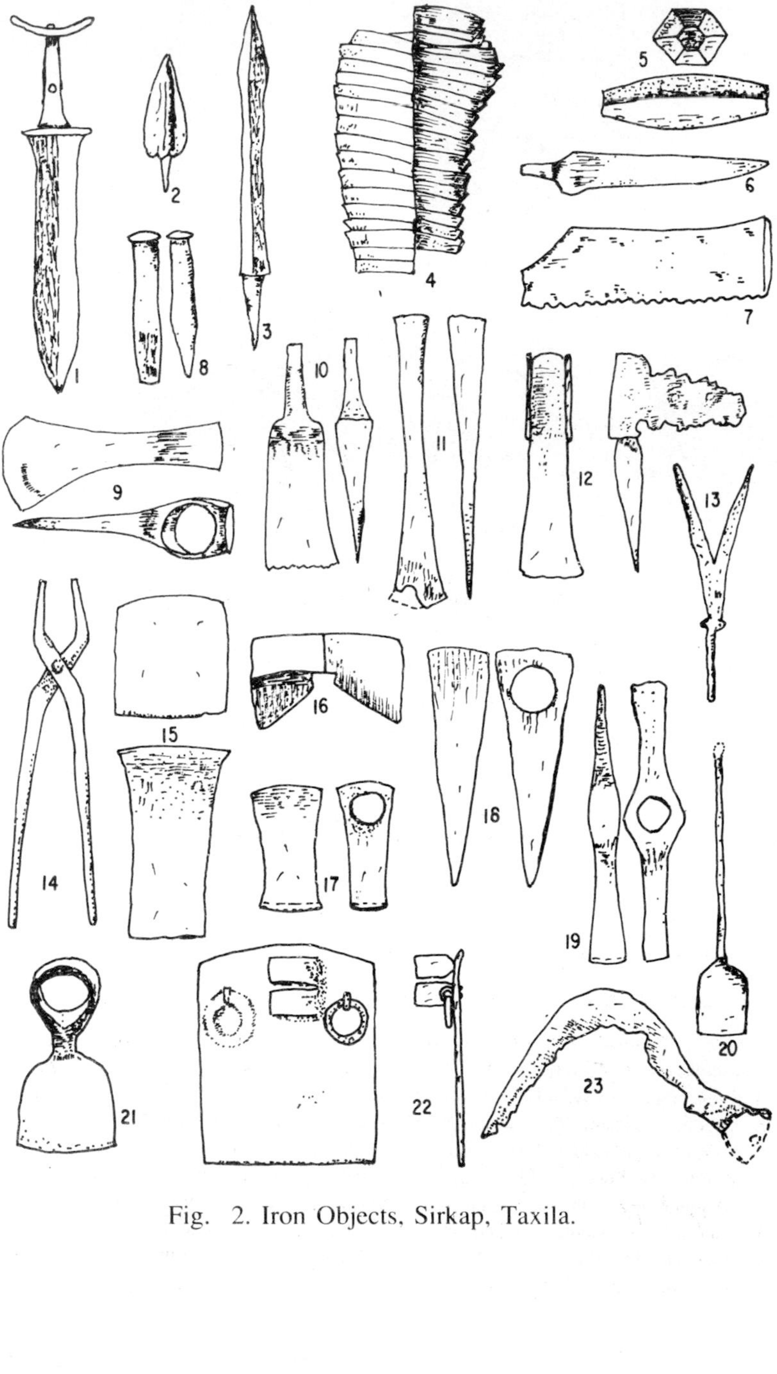

Fig. 2. Iron Objects, Sirkap, Taxila.

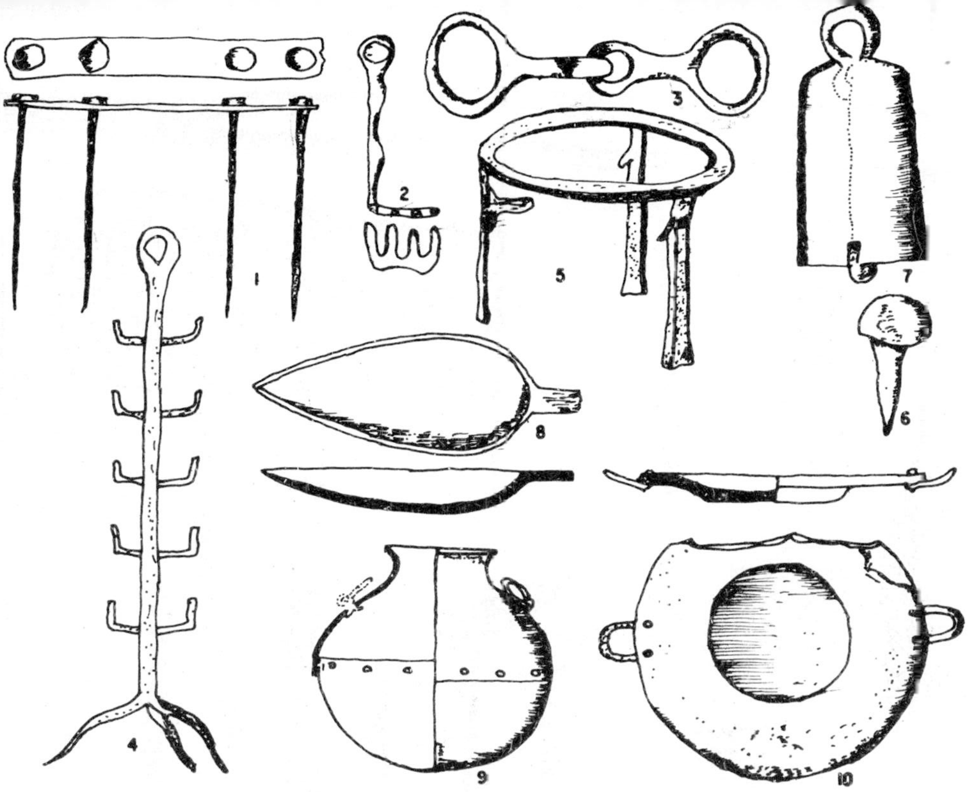

Fig. 3. Iron Objects, Sirkap, Taxila.

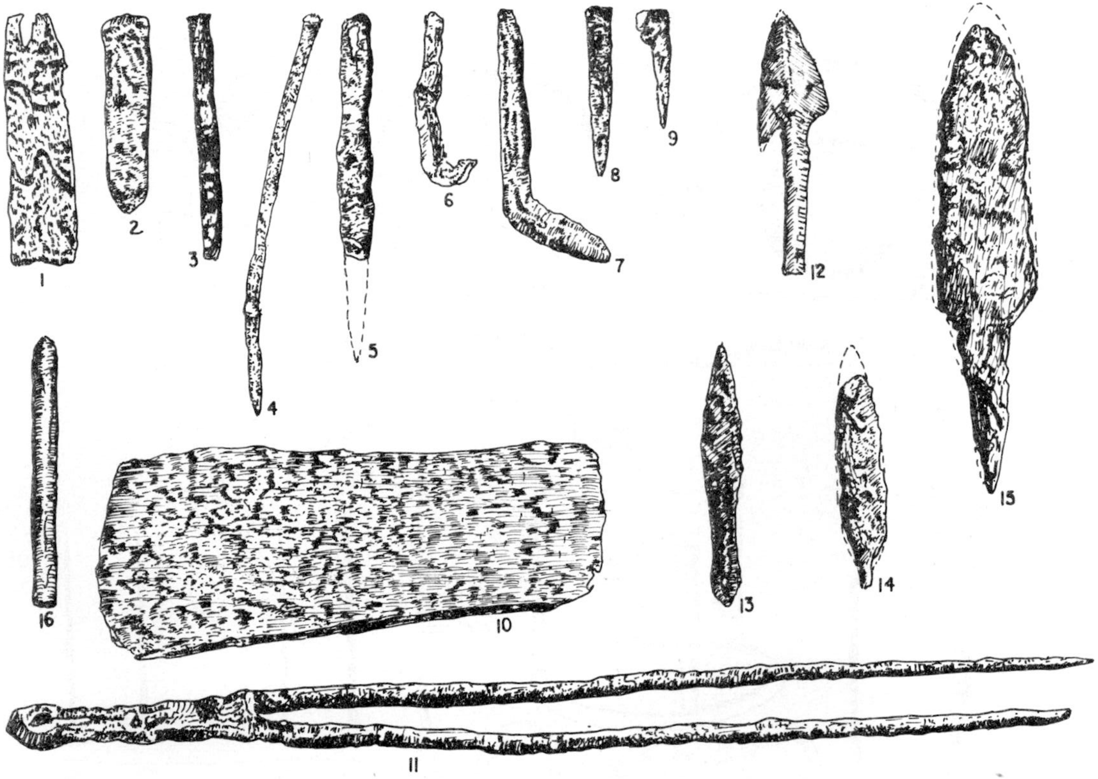

Fig. 4. Iron Objects, Period III, Atranjikhera.

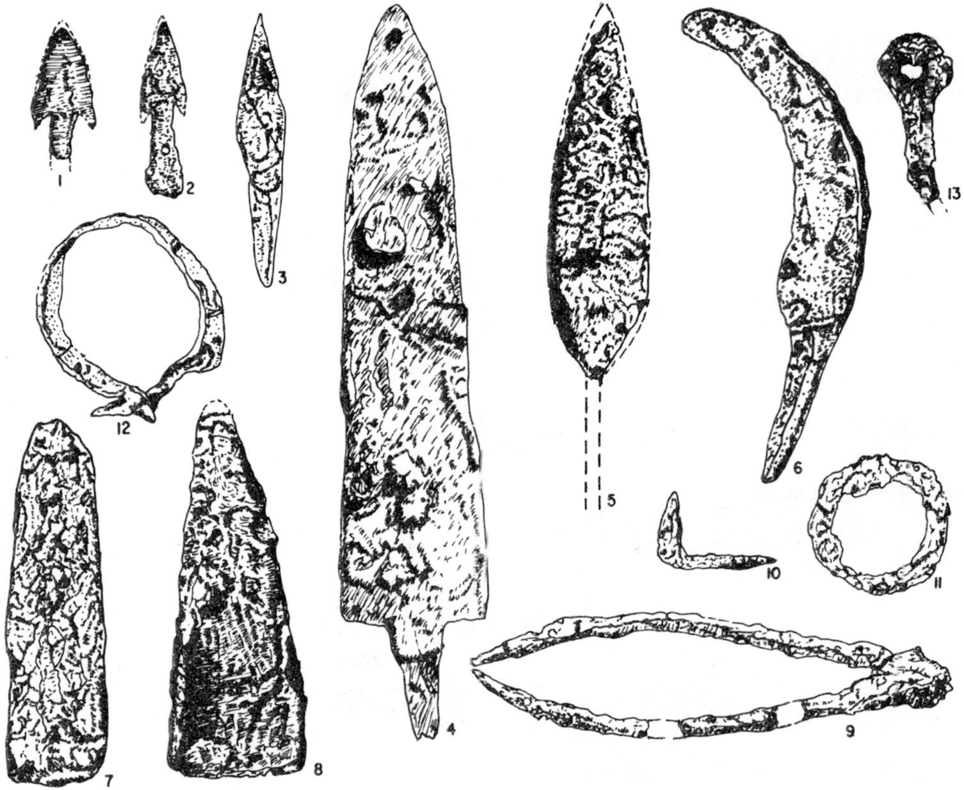

Fig. 5. Iron Objects, Period IV, Atranjikhera.

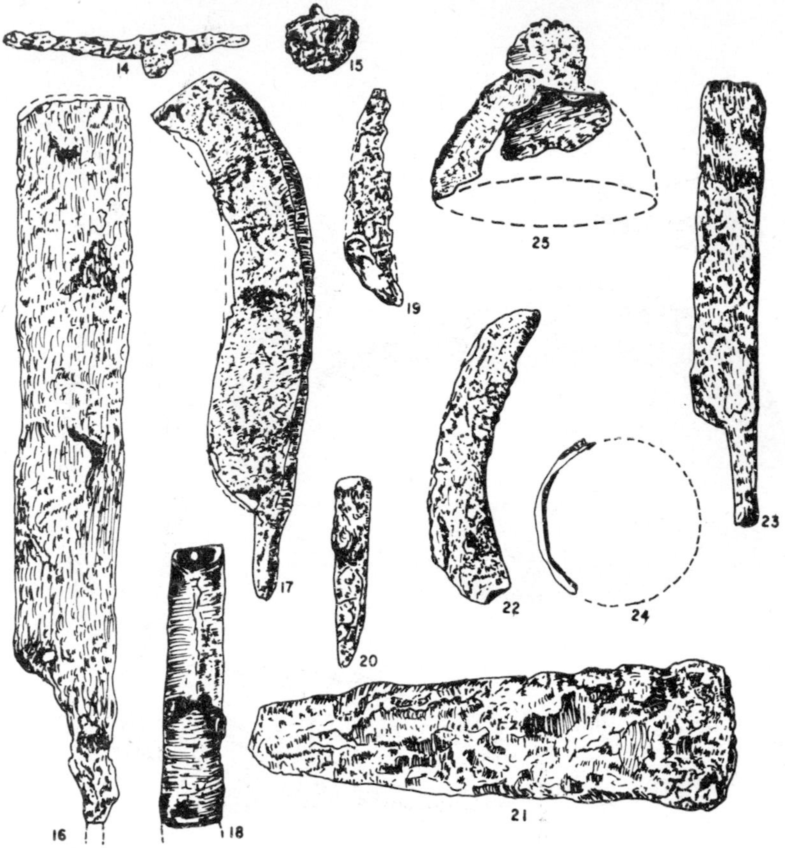

Fig. 6. Iron Objects, Period IV, Atranjikhera.

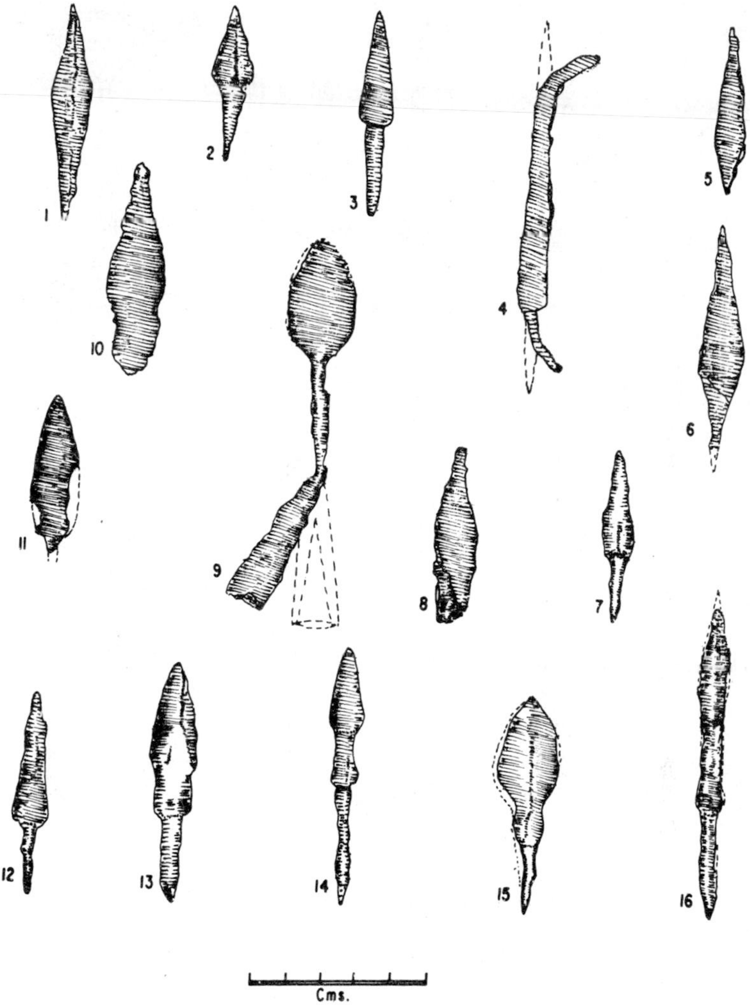

Fig. 7. Early Historic Arrowheads, Kausambi.

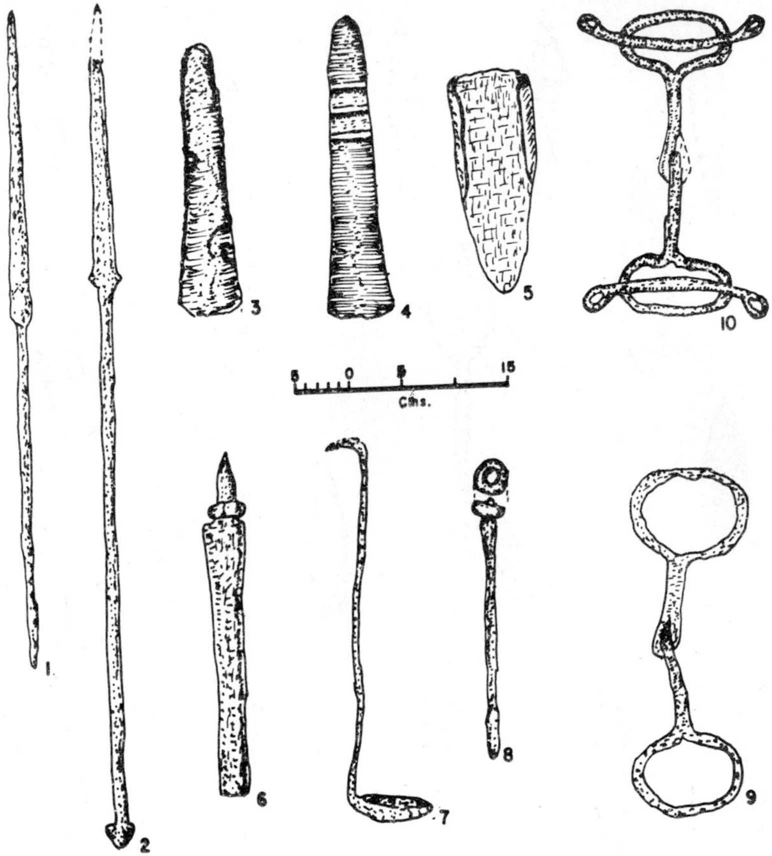

Fig. 8. Iron Objects, Mahurjhari

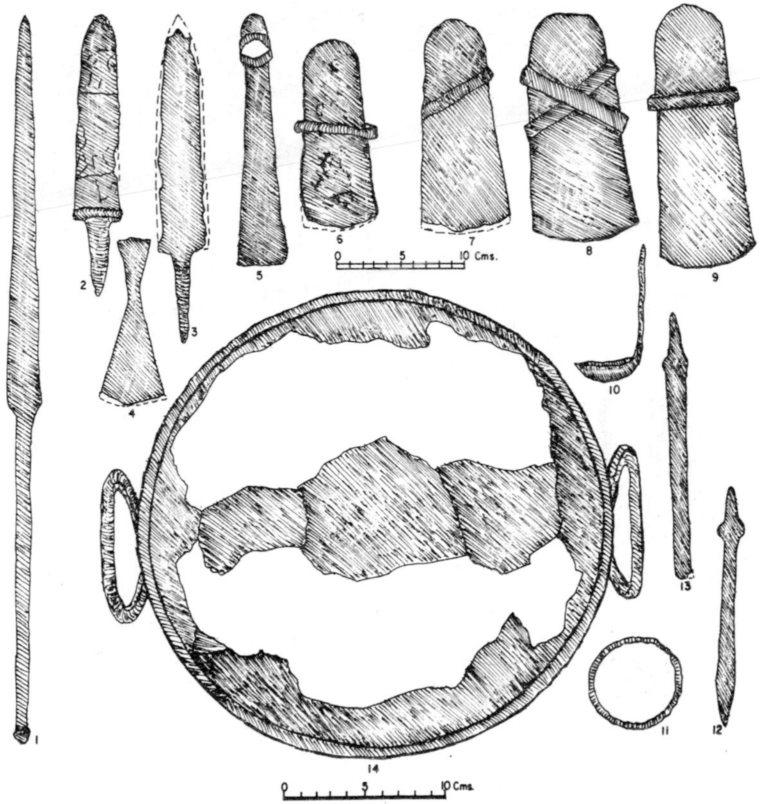

Fig. 9. Iron Objects, Takalghat-Khapa

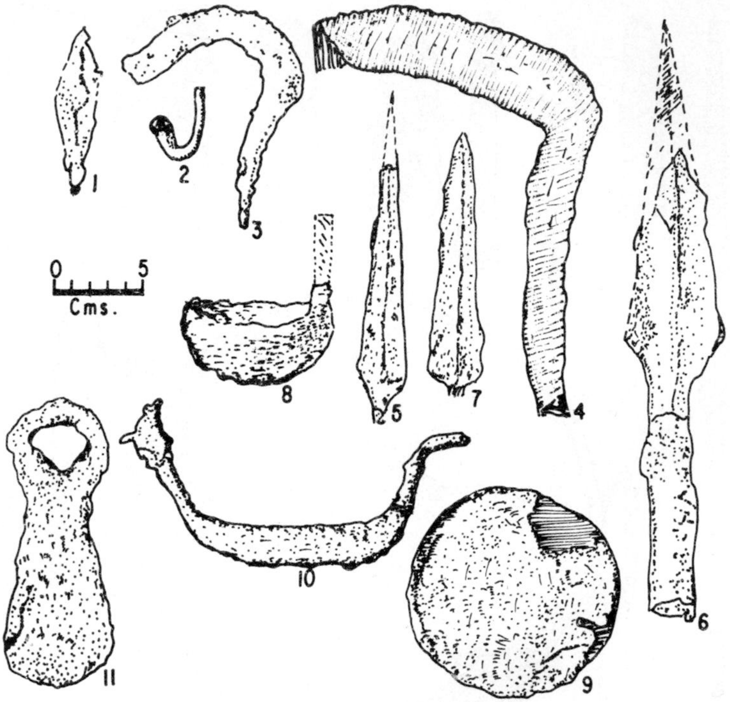

Fig. 10. Iron Objects, Nevasa

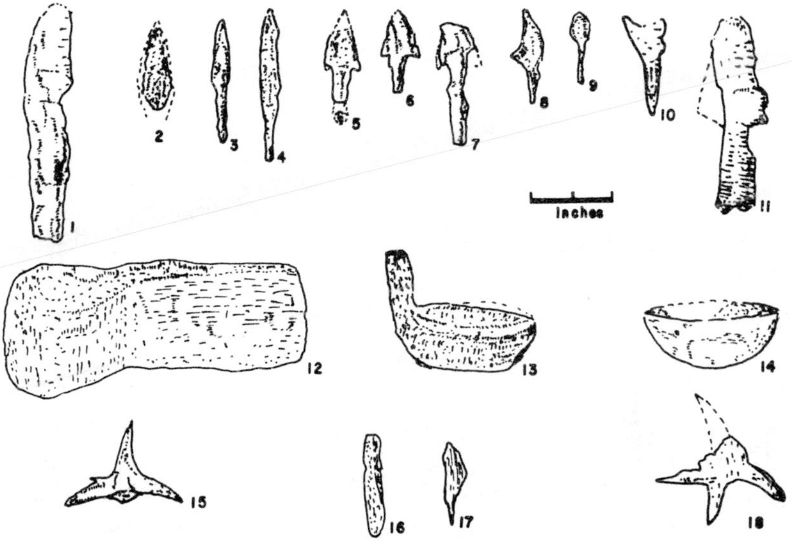

Fig. 11. Iron Objects, Nasik

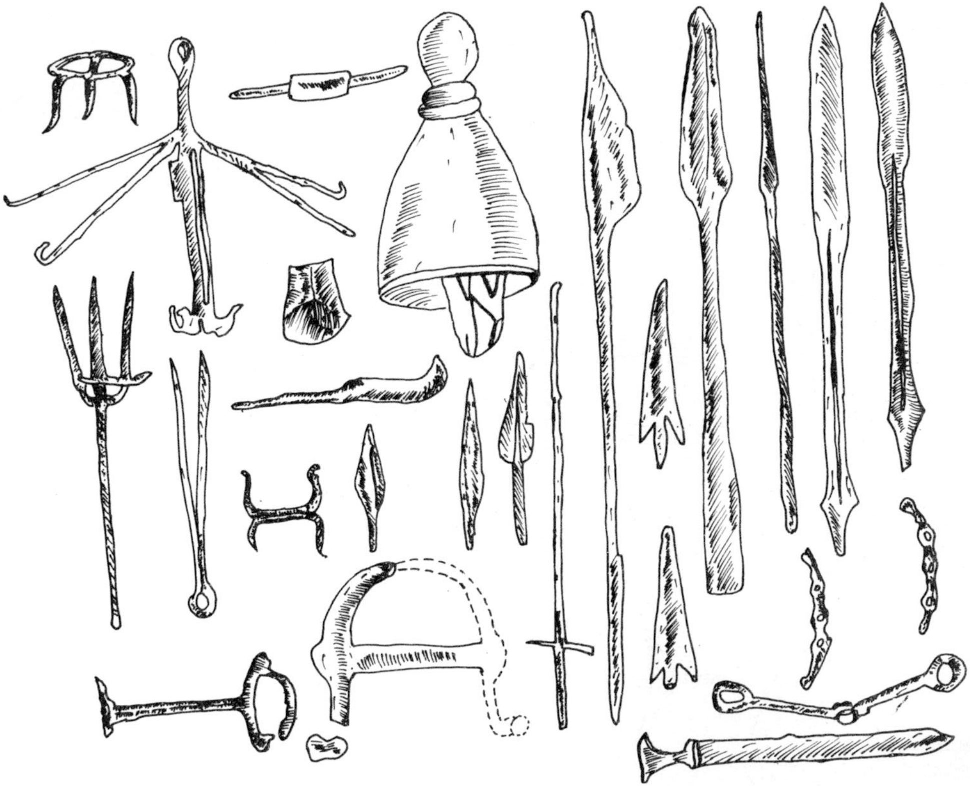

Fig. 12. Iron Objects, South-Indian Megaliths

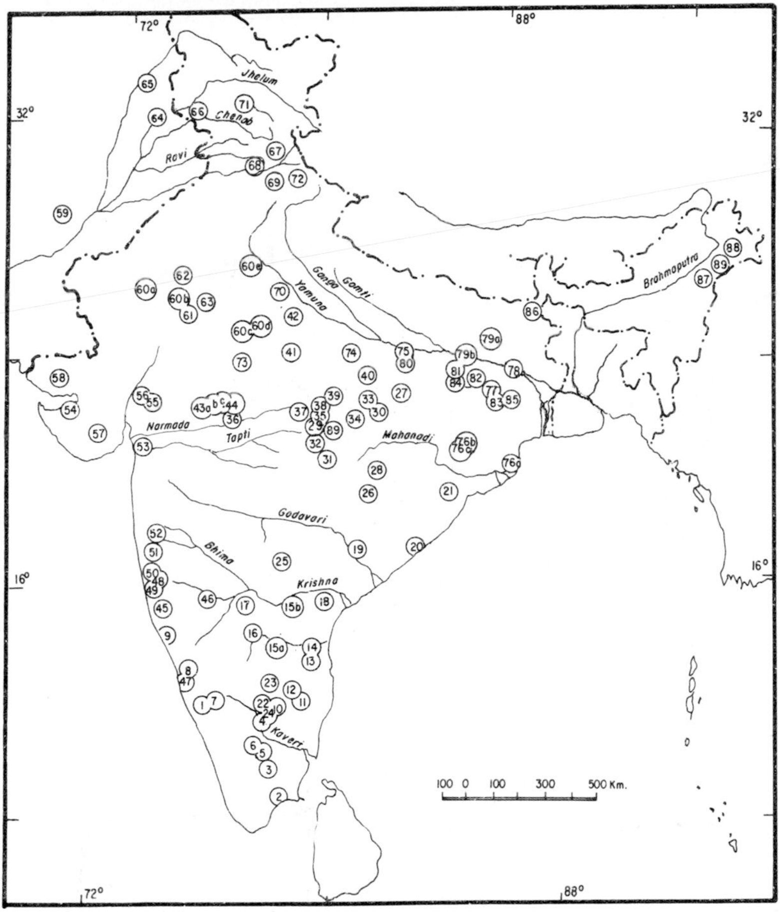

Fig. 13. Map Showing Distribution of Pre-industrial Iron-smelting.

1. Travancore, 2. Tinnevelli, 3. Madura, 4. Pudukottai, 5. Trichinopalli, 6. Coimbatore, 7. Nilgiri Hills, 8. Malabar, 9. South Kanara, 10. Salem, 11. South Arcot, 12. North Arcot, 13. Chingleput, 14. Nellore, 15a. Cuddapah, 15b. Kurnool, 16. Anantapur, 17. Bellary, 18. Krishna, 19. Godavari, 20. Vishakhapatnam, 21. Ganjam, 22. Ashtagram Division, Mysore, 23. Bangalore Division, 24. Nagar, Mysore, 25. Hyderabad, 26. Bastar, 27. Bilaspur, 28. Raipur, 29. Chanda, 30. Balaghat, 31. Bhandara, 32. Nagpur, 33. Mandla, 34. Seoni, 35. Chhindwara, 36. Nimar, 37. Hoshangabad, 38. Narsinghpur, 39. Jabalpur, 40. Rewa, 41. Bundelkhand, 42. Gwalior, 43a. Indore, 43b. Dhar, 43c. Chandgarh, Indore, 44. Ali Rajpur, Indore, 45. Dharwar, 46. Dharwar, 47. Kaladgi, 48. Belgaum, 49. Goa, 50. Sawantipur and Kolhapur, 51. Ratnagiri, 52. Satara, 53. Surat, 54. Rewakantha, 55. Panchmahals, 56. Kaira and Ahmedabad, 57. Kathiawar, 58. Cutch, 59. Sind, 60a. Narwar, 60b. Ajmer, 60c. Bundi, 60d. Kota, 60e. Bharatpur, 61. Mewar, 62. Jaipur, 63. Alwar, 64. Bannu, 65. Peshawar, 66. Kot Kirana Hills, 67. Kangra, 68. Mandi, 69. Simla Hills, 70. Gurgaon, 71. Kashmir, 72. Kumaun, 73. Lalitpur, 74. Banda, 75. Mirzapur, 76a. Cuttack, 76b. Talcher, 76c. Dhenkanal, 77. Burdwan, 78. Birbhum, 79a. Bhagalpur, 79b. Monghyr, 80. Gaya, 81. Hazaribagh, 82. Purulia/Manbhum, 83. Singhbhum, 84. Lohardaga, 85. Chhotanagpur Tributary States, 86. Darjeeling, 87. Khasi-Jaintia Hills, 88. Naga Hills, 89. Manipur

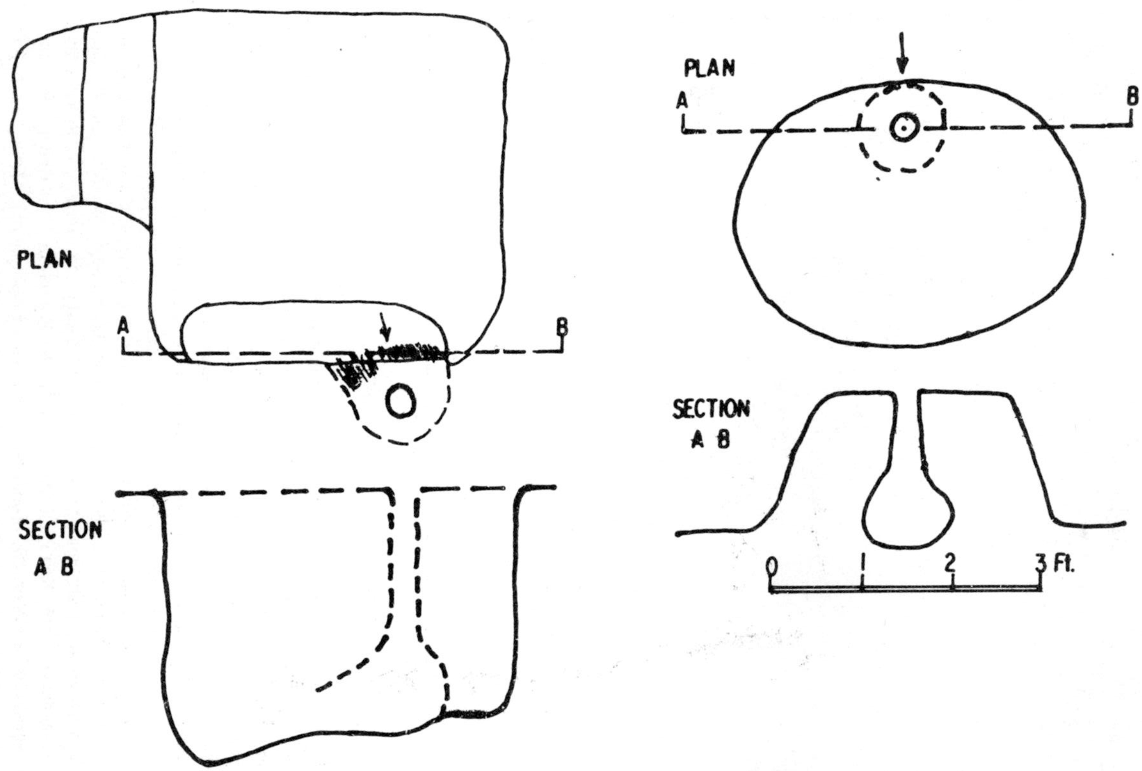

Fig. 14. Kamarjoda and Single Becha Furnaces, Singhbhum.

as denoting iron.[43] These metal objects are *kṣura* (razor, cf. 1.166.10, 10.28.9), *khāḍi* (Banerjee identifies it with *khāḍu* 'as it is now commonly called' or 'quoit rings' (cf. 6.16.40, 2.34.2, 10.38.1) and *asi* or *svadhiti* (axe, cf. 9.96.6, 3.8.11). The important references to the metallurgical activities are:

1. '*RV*, 10.81.3 where the act of welding has been positively referred to by the term *saṃdhamati* which Roth and Grassmann explain by *zussamen schewoissen* i.e., to weld together. Sāyaṇa explains the term *saṃdhamati* by *prerayati* or pressing into i.e., joining together (weld—to press or beat into intimate contact and permanent union, as two pieces of iron when heated almost to fusion).'[44]

2. 9.112.2: *jaratībhiḥ oṣadhibhiḥ parṇebhiḥ śakunānāṃ kārmāraḥ aśmabhiḥ dyurbhiḥ hiraṇyavantaṃ ichhati Indrāya indo pari srava.*[45] This passage has been cited by M. N. Banerjee (wrongly as *RV*, 10.72.2) and interpreted as evidence of steel manufacture in the Ṛgvedic period.[46]

3. 10.72.2: *eta saṃ karmārāḥ iva adhamat.*[47] According to Banerjee, 'karmāra and kārmāra of the Ṛgveda identified with his work *dham* signifies one working with bellows and fire and striking and hammering.'[48]

4. 4.2.17: 'performers of good works . . . have freed their birth from impurity, as (a smith heats) iron.' Sāyaṇācharya makes the meaning clear: *yathā karmārāḥ ayo bhaṣtreṇa dhamanti tadvat*—'as the smiths heat metal by the bellows'. M. N. Banerjee attached a lot of significance to this passage: 'That this fact of *purification* of ayas from ore at once dismisses any idea of identifying it with any kind of bronze is plainly evident.'[49] He adds as an explanation that in bronze-making two purified metals (i.e., copper and tin) are melted together to form an alloy. This process which is called melting is quite distinct from smelting.

Finally, it may be worthwhile to point out that Banerjee himself noted the following: 'In Yāska's synonyms for hiraṇya we have ayaḥ (Nirukta, Nighaṇṭu I, 2).'[50]
This obviously does not fit in with his main argument.

[43] M. N. Banerjee, 1927, 1929, 1932.
[44] M. N. Banerjee, 1927: 799.
[45] Max Muller, 1965, II: 259.
[46] M. N. Banerjee, 1929: 440.
[47] Max Muller, 1965, II: 329.
[48] M. N. Banerjee, 1927: 795–6.
[49] Ibid: 798.
[50] Ibid: 794.

The Later Vedas

In the *Sāmaveda*, the liturgical version of some of the Ṛgvedic hymns, *ayas* occurs only once, in the context of a wedge or club.[51]

In the *Taittirīya Saṃhitā* recension of the Black Yajurveda there are at least six references to *ayas* and at least one reference to smith. The references are: 1.2.11.f[52]—'That form of thine, O Agni, which rests in iron, which rests in silver, which rests in gold . . . ' (*yā te agne yāsaya rajāsaya harāsaya. . . .*); 1.8.12.m[53]—'gold-hued in the glowing of the dawns, bronze-pillared at the rising of the sun, O Varuṇa, O Mitra, mount your chariot seat' (*ayasthuṇaḥ*); 4.2.5.g[54]—'loosen ye this bond made of iron' (*ayasmayaṃ vandhanametat*); 4.6.7.i[55]—'golden his horns, iron his feet' (description of the sacrificial steed) (*ayaḥ asya pādāḥ*); 4.7.5—'may for me . . . gold, bronze, lead, tin, iron, copper . . . prosper through sacrifice'[56] (*hiraṇyaṃcha me ayaścha me sīsaṃcha me trapuścha me śyāmaṃcha me lohaṃcha me*)—the commentator explains: *śyāmam kṛsnāyasam loham kāṃsyatāmrādi*; 6.2.3—'the Asuras had three citadels; the lowest was of iron,then there was one of silver, then one of gold'[57] (*ayasmaya*); 4.5.4.n[58]—'homage to you potters, and to you, smiths, homage' (*karmārebhyaścha*).

It may be noted that *YV* 1.8.12.m and 4.6.7.i occur also in the *RV*. *YV* 4.7.5 is important, though the metal terminology is not quite clear. Besides gold, lead and tin (*hiraṇya, sīsa* and *trapu*), the terms used are *ayas, śyāmaṃ* and *lohaṃ*, translated as bronze, iron and copper respectively. According to the commentator, *śyāma* is black *ayas*, while *loha* is bell-metal, copper, etc. The epithet 'black' (*śyāma*) does make it iron and the use of this distinctive epithet in one instance makes the word *ayas* stand for bronze in this case.

In 2.11.5 of the recension known as the *Maitrāyanī Saṃhitā*, the relevant words which occur are *hiraṇya, sīsa, trapu, śyāmam, lohitāyasam*. These words figure in the *Vājasaneyī Saṃhitā* (18.13) as well.[59] Interestingly, *ayas* by itself is not used to mean bronze here;

[51] Benfey, 1848: 234.
[52] Keith, 1914: 30; Roer and Cowell, 1860–99, I: 388.
[53] Keith, 1914: 124; Roer and Cowell, 1860–99, II: 128.
[54] Keith, 1914: 314; Roer and Cowell, 1860–99, IV: 203.
[55] Keith, 1914: 376; Roer and Cowell, 1860–99, IV: 647.
[56] Keith, 1914: 381; Roer and Cowell, 1860–99, IV: 693.
[57] Keith, 1914: 504; Roer and Cowell, 1860–99, IV: 37.
[58] Keith, 1914: 357; Roer and Cowell, 1860–99, IV: 527.
[59] Schroeder, 1883, II: 142; Weber, 1852: 566–7.

the colour is denoted specifically. The specific denotation of bronze or copper by colour is found in another recension known as the *Kāṭhaka Saṃhitā* (18.10).[60] The word *śyāmenāyasā* occurs also in the *MS* 4.2.9, obviously in the sense of iron but in 4.2.13 of the *MS* *ayas (āyasī)* occurs without any epithet,[61] though this does not necessarily mean that here it refers to bronze. However, despite this somewhat confused meaning of *ayas* one can positively infer that, by the time of the *YV*, iron was a known metal. There is no particular way of inferring the extent of its use but one may note that there are references in the *Taittirīya Saṃhitā* to ploughs drawn by 6 or even 12 oxen.[62] There is no reason why these ploughs could not have been bronze-tipped or even wooden but the use of iron also seems to be an equally fair possibility.

There are quite a few references to plain, unqualified *ayas* in the *Atharvaveda (AV)*. Some of these references are: 8.3.2: 'do thou, of iron tusks, O Jātadevas' (text: *ayodaṃṣṭra*);[63] 11.10.3: 'iron-mouthed—let the flesh-eaters . . . fasten on our enemies' (text: *ayomukhāḥ*);[64] 5.28.1: 'with an amulet of three metals nine breaths with nine he combines in order to length of life for a hundred autumns; in the yellow three, in silver three, in iron three—enveloped with fervour' (text: *ayasi*);[65] 5.28.5: 'let Agni rescue thee in accord with the iron' (text: *ayasā*);[66] 5.28.9: 'from the earth let that made of iron protect thee' (text: *ayasmayam*);[67] 7.120.1: 'with a hook of metal we attach thee to him who hates us' (text: *ayasmayena*);[68] 4.37.8: 'Indra's missiles, a hundred spears of iron' (text: *ayasmayīḥ*);[69] 6.63.2: 'unfasten the bond-fetters of iron' (text: *ayomayān*);[70] and 6.63.3: 'iron post' (text: *ayasmaye*).[71] Apart from these and a few other similar references (cf. 8.13.2, 6.84.4, 19.66.1) the meaning of iron is made specific only twice: 9.5.4: *śyāmena asinā*—'by the knife of dark metal', and 11.3.7; *śyāmāyo asya māṃsaṃ lohitamasya lohitam*—

[60] Schroeder, 1900.
[61] Schroeder, 1886: 32, 36.
[62] Keith, 1914: 408; the *TS*, 5.2.5.
[63] Whitney, 1905: 481; Vishva Vandhu (cited hereafter as Vish.), 1961–4: 1087.
[64] Whitney, 1905: 656; Vish., 1961–4: 1427.
[65] Whitney, 1905: 272; Vish., 1961–4: 624.
[66] Whitney, 1905: 272; Vish., 1961–4: 625.
[67] Whitney, 1905: 273; Vish., 1961–4: 626.
[68] Whitney, 1905: 464; Vish., 1961–4: 1054.
[69] Whitney, 1905: 213; Vish., 1961–4: 546.
[70] Whitney, 1905: 328; Vish., 1961–4: 738.
[71] Whitney, 1905: 328; Vish., 1961–4: 738.

'dark metal its flesh, red its blood'.[72] In 10.1.20 there may be a reference to a magical property of iron: 'there are knives of good metal in our house . . . O witchcraft, go away from here' (*svayasā asayāḥ santi*).[73] There is a reference to *karmārāḥ*—smiths—in 3.5.6[74] while the verb expression *dhamantar* occurring in 18.3.22, is translated as 'forging'.[75] In 11.9.1 there are references to arrows, bows, swords and axes,[76] and in 12.5.20 there is the adjective 'keen-edged'.[77] An indirect evidence of the extent of the use of iron seems to be provided by 10.6.2–3 *AV* 10.6.2 speaks of an amulet born of the ploughshare and in 10.6.3. there is the following expression regarding the amulet: 'the skilful smith hath smitten thee away with the hand by a knife'.[78] The necessity of the reference to a smith in the context of the manufacture of an amulet out of a ploughshare seems to suggest unequivocally that the ploughshare was of metal and, considering the fact that iron was positively known in the *AV*, the metal ploughshare in this case might well have been of iron.

The Brāhmaṇas

Among the *Brāhmaṇas* the Ṛgvedic *Brāhmaṇas* like the *Aitareya*[79] and the *Kauṣitakī*[80] do not specifically mention iron or even unspecified *ayas* while the *Taittirīya Brāhmaṇa* of the Black *YV* seems to mention *ayas* (3.4.10: *ayastāpam*)[81] only once, but in an unspecified manner. The references in the *Śatapatha Brāhmaṇa* of the same *Veda* are, however, important and deserve detailed consideration.

The general references like those in 3-4.4.3,[82] 12-7.1.7,[83] 5-4.1.2.,[84] etc. are not particularly revealing. In 5-4.1.1[85] copper or bronze is specifically mentioned as *lohāyasa* and, in 6-1.3.5,[86] there

[72] Whitney, 1905: 533; Vish., 1961–4: 1181; Whitney, 1905: 626; Vish., 1961–4: 1330.

[73] Whitney, 1905: 565; Vish., 1961–4: 1219.

[74] Whitney, 1905: 92; Vish., 1961–4: 294.

[75] Whitney, 1905: 855; Vish., 1961–4: 1715.

[76] Ibid.

[77] Ibid.

[78] Vish., 1961–4: 1240.

[79] Keith, 1920.

[80] Ibid.

[81] Mitra, 1870: 355.

[82] Eggeling, 1885–1900, XXVI: 105.

[83] Ibid, XLIV: 215.

[84] Ibid, XLI: 328.

[85] Ibid: 90.

[86] Ibid: 158.

is obviously a reference to metallurgy, which in translation reads 'whence ore much smelted comes, as it were, to have the appearance of gold'.

What is particularly significant is that in 13-2.2.16-19[87] iron has been associated with the peasantry or the common people. The context is the ritual of the horse-sacrifice and in 13-2.2.16 one reads: 'The slaughtering-knife of the horse is made of gold, those of the *paryaṅgas* (literally this term means 'body-encircling', but in this case it denotes 15 other animals which are directly related to the horse in the ritual and also to be sacrificed) of iron.' 13-2.2.17 has 'gold is a form of the nobility' (the horse is associated with the nobility). 13-2.2.18 states: 'And as to why there are copper (knives) for the *paryaṅgas*,—even as the non-royal king-makers, the heralds, and headmen are to the king, so these *paryaṅgas* are to the horse; and so indeed is this—to wit, copper to gold . . .' 13-2.2.19 continues: 'and as to why there are iron ones for the others—the other animals, indeed, are the peasantry, and this—to wit, iron—is a form of the peasantry . . .' (*sadāyasa itareṣām*). The original of the relevant portion in 13-2.2.16 is: *hiraṇyamayośvasya śāso bhavati lohamayāḥ paryaṅgayānamāyasā itareṣām*.

The association of iron with the common people is also made evident in 13-3.4.5 where the third oblation of blood after the horse sacrifice is supposed to be given 'in an iron bowl; for the people (subjects) are of iron' (*ayasmayena charuṇā tṛtīyām āhutiṃ juhoti āyasya vai prajā*).[88]

The Upaniṣads

In the philosophical abstractions of the *Upaniṣads* one need not expect to find any reference to iron except as a metaphor. There are several references of this kind in some of the principal *Upaniṣads*. *Chhāndogya* 4.17.7:[89] *tad yathā lavaṇena suvarṇaṃ saṃdadhyāt suvarṇena rajatam rajatena trapu trapuṇā sīsam sīsena loham lohena dāru dāru charmaṇā*—'just as one would bind together gold with (borax) salt, silver with gold, tin with silver, lead with tin, iron with lead, wood with iron or wood with leather.' From the context it does not become very clear whether iron or copper/bronze is indicated. *Chhāndogya* 6.1.5.:[90] *yathā saumya ekena lohamaṇinā sarvaṃ*

[87] *Ibid, XLIV: 303–4; Weber, 1855: 966.*
[88] *Ibid: 339; Weber, 1855: 976.*
[89] *Radhakrishnan, 1953: 420.*
[90] *Ibid: 447.*

lohamayaṃ vijñātaṃ syāt vāchārambhaṇam vikāro nāma-dheyam lohamity eva satyam—'just as, my dear, by one nugget of gold, all that is made of gold becomes known, the modification being only a name arising from speech, while the truth is that it is just gold.' According to the translator, S. Radhakrishnan,[91] *loha* originally meant iron or copper but later was used for gold or any metal.

Chhāndogya 6.1.6:[92] *ekena nakha-nṛkṛntanena sarvaṃ kārṣṇāyasaṃ vijñātam syāt*—'by one pair of nail-scissors all that is made of iron is known.'

Maitrī 6.27:[93] *achireṇaiti bhumāv ayaspiṇḍam nihitam yathā'chireṇaiti bhūmitvam, mṛdvat saṃsthnam ayaspiṇḍam yathāgnyayaskārādayo nābhibhavanti*—'even as a ball of iron that is hidden in the earth passes speedily into the condition of earthiness. As fire and blacksmiths and the like do not trouble about the ball of iron that is in the condition of earth . . .'

Maitrī 3.3.:[94] *yathāgnināyaspiṇḍo vābhibhūtaḥ kartṛbhir hanyamāno nānātvam upaity evam* . . . *yathāyaspiṇḍe hanyamāne nāgnir abhibhūyaty evam*—'even as a ball of iron, overcome by fire and beaten by workmen takes many forms . . . as when a ball of iron is being beaten, the fire is not overcome.'

There are also a few other references: *paraśuṁ taptam* (heated axe, *Chhāndogya* 6.16.1[95]), *kṣurasya dhārā* (edge of a razor, *Kaṭha* 1.3.14[96]), *tad yathā kṣurāh kṣura dhāne'vopahito* (just as razors might be hidden in razor-case, *Kauṣitakī* 4.20[97]).

The Vedic Kalpasūtras

Whether or not one accepts Ram Gopal's dating of this body of literature as between c. 800 B.C. and c. 500 B.C. (he argues that Pāṇini was familiar with the Sūtra literature), his references to the metallurgical activities are relevant to our context. 'Blacksmiths and goldsmiths are separately mentioned in the *Baudhāyana Śrauta Sūtra* (XV. 13) . . . Blacksmiths manufactured articles of iron, copper and bell-metal. Ploughshares, spades, sickles, needles, knives, utensils, and razors were some of the important articles manufactured by blacksmiths.'[98]

[91] Ibid.
[92] Ibid.
[93] Ibid: 836–7.
[94] Ibid: 806–7.
[95] Ibid: 466.
[96] Ibid: 628.
[97] Ibid: 791.
[98] Gopal, 1959.

Pāṇini and Patañjali

The cultural material in Pāṇini's *Aṣṭādhyāyī* has been studied by V. S. Agrawala.[99] He has listed the following references to metals: gold (*hiraṇya* or *jātarūpa*), silver (*rajata*), lead (sīsa), tin (*trapu*), bell-metal (*kāṃsya*), copper/bronze (*loha*) and iron (*ayas*).[100] Terminologically, what is significant is that Pāṇini (5.4 94[101]) takes *ayas* both as a *jāti* (genus) and as *saṃjña* (species)—*anośmāyaḥ sarasām jātisaṃjñāyoḥ*. According to Agrawala[102] this is 'illustrated by the *Kaśikā* as *kālāyasa* (iron) and *lohitāyasa* (copper) respectively'. One does not understand how iron and copper can be regarded as a genus and species. A more logical explanation seems to be that, here, by *jāti* Pāṇini means *ayas* as a generic name for both iron and copper or even metal in general while by *saṃjñā* he refers to instances when *ayas* means only iron. S.M. Katre in his *Dictionary of Pāṇini*[103] lists some of the references to *ayas* in the *Aṣṭādhyāyī: ayaḥ-śūla* (5.2.76), *ayas* (3.3.82, 5.4.94, 8.3.49), *ayas-kaṁsa* (8.3.46, an iron goblet), *ayas-karṇī* (8.3.46, a woman having hard ears), *ayas-kāma, ayas-kāra* (8.3.46, a blacksmith), *ayas-kumbha* (8.3.46, an iron pot or boiler), *ayas-kuśā* (8.3.46, a rope partly consisting of iron), *ayas-pātra* (8.3.46, iron pot), *ayas-maya* (1.4.20, made of iron or metal), *ayo-ghana* (3.3.82, an iron hammer), *ayo-datī* (5.4.143, having teeth like iron), etc. *Ayovikāra kuśi* has been interpreted by Agrawala as iron ploughshare. He also refers to a class of ascetics, *ayaḥśūlikas*, 'who flourished by the method of *ayaḥśūla* or iron spikes' (i.e., by brandishing their iron spikes or tridents). A few related terms also occur in the *Mahābhāṣya* of Patanjali.[104] These are *ayaskāra, ayaskumbha*, etc.

The Epics

The references are generally in the context of arms and armour, many of which were of iron.[105] For instance, in the *Mahābhārata* (*Mbh*) Kṛshṇa's wheel was made of iron, sharp-edged and revolving—*paribhramantaṃ, tīkṣṇadhāraṃ* and *ayasmayain*. There is reference

[99] Agrawala, 1953.
[100] Ibid: 231.
[101] Sastrigal, 1912: 142.
[102] Agrawala, 1953: 231.
[103] Katre, 1968.
[104] Kielhorn, 1962; 4.1.1; 3.1.7; also Puri, 1968.
[105] The evidence of arms and armour, many of which were of iron, in the epics has been collated by S. D. Singh, 1965 (pp. 105–18).

to black *ayas* in the *Droṇa-parva* of the *Mbh.*[106] The words commonly used to denote a club or mace are *gadā, muṣala* and *parigha*. There are references to these being made of *ayas* in the *Mbh.*[107] The *Araṇyakāṇḍa* of the *Rāmāyaṇa* speaks of a *parigha* with iron prickles.[108]

The Arthaśāstra

By the time the text of the *Arthaśāstra* was laid down, the art of metallurgy was a highly organized affair, the responsibility of the organization resting on the Directors of various government departments, one of whom was the Director of Metals. According to the text 'the director of metals should establish factories for copper, lead, tin, vaikrintaka, brass, steel, bronze, bell-metal and iron, also (establish) trade in metal-ware' (*lohādhyakṣaḥ tāmrasīsatrapuvaikṛintakārakūṭavṛittaka ṃsatālalohakarmāntān kārayet. Lohabhānḍavyavahāraṃ cha*).[109]

For the earlier stage of ore-extraction there was a Director of Mining also, who among other things was supposed to 'inspect an old mine by the marks of dross, crucibles, coal and ashes, or a new mine where there are ores in the earth'. The scope of the mines is expressed clearly: 'gold, silver, diamonds, gems, pearls, corals, conchshells, metals, salt and ores derived from the earth, rocks and liquids—these constitute mines' (*suvarṇarajatavajramaṇimuktāpravālaśaṃkhalohalavanabhūmi prastararasadhātavaḥ khaniḥ*).[110] Partly, the metals also came under the Director of Forest Produce. The familiarity of the *Arthaśāstra* with iron deserves no special emphasis, even though the direct references to specific iron objects seem to be somewhat limited: iron rod (4.1.62–*ayaśśula*;[111] 2.19.10—weights of iron (*ayomayāni*);[112]; 13.2.23—iron clubs and pestles (*śaktimuṣalānyayomayāni*),[113] etc. What possibly is more interesting is that in 4.4.20 and 2.15.60 some idea may be had of the technical equipment of the contemporary blacksmiths and metalsmiths in gen

[106] De, 1959: 167; cf. 7.28.4.
[107] 3.234.21; 6.66.18; 1.17.6.
[108] The Rām., 26.10.11; Parab, 1902.
[109] Kangle, 1963: 124; Shama Sastri (cited hereafter as Sastri), 1919: 94.
[110] Kangle, 1963: 87; Sastri, 1919: 60.
[111] Kangle, 1963: 299; Sastri, 1919: 204.
[112] Kangle, 1963: 154; Sastri, 1919: 103.
[113] Kangle, 1963: 556; Sastri, 1919: 398.

eral. 4.4.20 occurs in the context of the operational mode of royal spies inside the kingdom, and it explicitly suggests a way of exposing a manufacturer of counterfeit coins. The text reads, 'if he considers any one as an utterer of false coins, (being) a frequent purchaser of various metals and acids, of coals, bellows, pincers, vices, anvils, dies, chisels and crucibles, with indications of hands and clothes smeared with soot, ashes and smoke, (and being) possessed of blacksmith's tools, a secret agent should expose him by insinuating himself into his confidence as a pupil and by carrying on dealings with him' (*yaṃ vā nānālohakṣārāṇāṃ aṅgārabhastrāsaṃdaṃśamūṣikādhikaraṇiviṭanka mūṣāṇāmabhīkṣṇaṃ kretāraṃ mūṣībhasamadhūmadigdhahastabastraliṅgaṃ karmāropakaraṇasaṃsargaṃ kūtarūpakārakam manyet*).[114]

2.15.60 says in the context of the duties of the Superintendent of Magazines that 'he should cause charcoal and husks to be taken to metal workshops and for plastering walls' (*aṅgārān tuṣān lohakarmāntabhittilepyanāṃ hārayet*).[115] 7.3.7. suggests an equally easy familiarity with the working of metalsmiths: 'for heat is the means of joining together, metal that is not heated does not become joined with metal' (*tejo hi saṇdhanakaraṇaṃ nātaptaṃ lohaṃ lohena sandhatta iti*).[116]

What, however, is particularly interesting in the *Arthaśāstra* is its metal terminology. First, the generic name of all metals is *loha*. *Lohādhyakṣa* (2.12.23) is the name of the Director of Metals. *Loha* in this sense occurs again and again:

2.17.14 — *kālāyasatāmravṛittakāṃsyasīsatrapuvaikṛintakārakūṭāni lohāni*—'iron, copper, steel, bronze, lead, tin, vaikrintaka and brass (constituting the group of) metals.'[117] In B. Shama Sastri's translation, *vaikrintaka* is mercury.[118]

2.6.4—*loha* in the context of the mines.

2.22.6—*lohavarṇadhātunāṃ*—metals of various kinds and ores.[119]

1.21.7—*lohamaṇimayānāṃ*—metals and gems.[120]

5.2.9—*lohapaṇyāḥ*—metals.[121]

[114] Kangle, 1963: 308; Sastri, 1919: 212.
[115] Kangle, 1963: 144; Sastri, 1919: 97.
[116] Kangle, 1963: 379; Sastri, 1919: 269.
[117] Kangle, 1963: 149; Sastri, 1919: 100.
[118] Sastri, 1929: 109.
[119] Kangle, 1963: 167; Sastri, 1919: 112.
[120] Kangle, 1963: 59; Sastri, 1919: 43.
[121] Kangle, 1963: 345; Sastri, 1919: 243.

2.21.22—*śastravarmakavachaloharatharatnadhānya paśūnām*—'weapons, armours, coats of mail, metals, chariots, jewels, grains and cattle.'[122]
2.19.13—*lohapalāt*—a ball of metal.[123]
2.19.24—*loha*—metal.[124]
3.19.12—*loha*—metal.[125]
There are a few more references of this kind, but *loha* has possibly been used in other specific senses also:
2.14.25—*lohapiṇḍavālukābhiḥ*—sand lumps containing 'gold';[126] 5.1.33, 13.2.29—*lohamuṣalaiḥ*—'iron' clubs;[127] 2.18.16—*lohajālikāpaṭṭakavachasūtrakaṃ*—a coat of mail of 'iron' rings or plates,[128] etc. In all these contexts *loha* may simply mean 'metal', but the possibility of its being gold in one case and iron in the rest can also not be denied.
Loha again may not be the only term used for metals. In 2.22.6 (*lohavarṇadhātunāṃ*) *dhātu* has been taken as ores and *loha* as metal[129] but in 2.22.10 (*dhātupaṇyādāneṣu*) *dhātu* has been translated as metal.[130]

Secondly, as has already been noted, *ayas* has been used to denote iron. But *kālāyasa* also has been used, at least thrice: 2.17.14—*kālāyasatāmra* . . .; 3.17.7-8—*kālāyasa* . . . *tāmravrittakāṃsakācha* (iron, copper, steel, bronze, glass;[131] 2.13.55—*kālāyasa*. The third word used for iron is *tīkṣṇa*: cf. description of the iron ore—*kurūmvaḥ pāṇḍuro hitassinduvārapuṣpavarṇo vatīkṣṇadhātu* (2.12.15)—'that which is made up mostly of smooth stones, is whitish-red or of the colour of sinduvāra flower is iron ore'.[132] The term also occurs in 2.12.24 and 2.13.48.[133] *Karmāra* (14.1.35) may denote blacksmiths

[122] Kangle, 1963: 164; Sastri, 1919: 111.
[123] Kangle, 1963: 154; Sastri, 1919: 103.
[124] Kangle, 1963: 156; Sastri, 1919: 104.
[125] Kangle, 1963: 288; Sastri, 1919: 195.
[126] Kangle, 1963: 136; Sastri, 1919: 91.
[127] For 5.1.33, Kangle, 1963: 341 and Sastri, 1919: 240; for 13.2.29, Kangle, 1963: 556 and Sastri, 1919: 398.
[128] Kangle, 1963: 152; Sastri, 1919: 102.
[129] Kangle, 1963: 167; Sastri, 1919: 112.
[130] Kangle, 1963: 167; Sastri, 1919: 113.
[131] For 2.17.14, Kangle, 1963: 149 and Sastri, 1919: 100; for 3.17.7–8, Kangle, 1963: 285 and Sastri, 1919: 192.
[132] For 2.13.55, Kangle, 1963: 132, Sastri, 1919; 89; for 2.12.15, Kangle, 1963: 123, Sastri, 1919: 83.
[133] Kangle, 1963: 124; Sastri, 1919: 84.

and *kārmārika* (2.3.35) 'produce of blacksmiths'.[134]

Thirdly, there is new word, not used earlier: *vṛtta* (2.17.14, 3.17.7-8, 4.1.35) translated as 'steel' by Kangle but as 'bronze' by Shamasastry.[135]

The Sanskrit Lexical Literature

We may refer to only two of the Sanskrit lexicons—*Amarakoṣa* which need not be earlier than the sixth century A.D. and *Abhidhāna-ratnamālā* which could be composed towards the end of the eleventh century. In the *śloka* 18 of *Amarakoṣa* the following terms occur: *loha, tīkṣṇa* and *kālāyasa*. The term *loha* denotes metal in *Amarakoṣa*. The commentary on the succeeding *śloka* 19 makes it clear: *sarvamapi taijasaṃ suvarṇarajatādikaṃ lohamityuchchate—suvarṇaṃ rajataṃ tāmraṃ rītiḥ kāṃsyaṃ tathā trapu. Sīsaṃ kālāyasaṃ chaivamaṣtau lohāṃ chakṣata iti.*[136]

The terms, *loha*, *kālāyasa* and *tīkṣṇa*, also occur in the *Abhidhāna-ratnamālā* (2.16).[137]

In *Indian Epigraphical Glossary*, D. C. Sircar points out that in the inscriptions *loha* is used in the sense of metal. The term *lohāra* is a modification of *lohakāra*, a blacksmith. *Loha-vāṇija* denotes an iron-monger whereas *lohika-kāruka* means a worker in metal.[138]

The Sanskrit Chemical and Alchemical Literature

Varāhamihira's *Bṛhatsaṃhitā* (fifth-sixth centuries A.D.) has a chapter entitled *khaḍgalakṣaṇam* or signs of swords. This chapter contains some empirical recipes for the hardening of steel. 'Plunge the steel, red-hot, into a solution of plaintain ashes in whey, keep standing for twenty-four hours; then sharpen on the lathe.'[139] 'Make a paste with the juice of the plant *arka* (*Calotropis gigantea*), the gelatine from the horn of the sheep, and the dung of the pigeon and the mouse; apply it to the steel after rubbing the latter with sesame oil. Plunge the steel thus treated into fire, and when it is red hot, sprinkle on it water or the milk of horse (camel or goat), or *ghee* (clarified butter), or blood, or fat, or bile. Then sharpen on the lathe.'[140]

[134] For 14.1.35, Kangle, 1963: 576, Sastri, 1919: 413; for 2.3.35, Kangle, 1963: 77, Sastri, 1919: 54.

[135] cf. Kangle, 1963: 149; Sastri, 1929: 109.

[136] Thatte, 1877: 235.

[137] Aufrecht, 1861: 19.

[138] Sircar, 1966: 172.

[139] Cited in P. Ray, 1956: 103.

[140] Ibid.

P. C. Ray, who was among the first to discuss Indian literary data on iron in some detail, drew attention to a text called *Rasaratna-samuchchaya* which he placed between A.D. 1300 and 1550. Though it is very much later than the chosen time limit of the present volume, its significance is too great to be entirely omitted. In the section on iron this text begins by saying that there are three kinds of iron: *muṇḍa, tīkṣṇa* and *kānta*. *Muṇḍa* is of three varieties: *mṛdu, kunṭha* and *kaḍāra*. *Mṛdu* melts easily, does not break, and is not glossy. *Kunṭha* expands with difficulty when struck with a hammer. *Kaḍāra* breaks under the hammer and has a black fracture. Ray classified *muṇḍa* as wrought iron. *Tīkṣṇa* is steel (Ray suggested cast iron as well) and is supposed to be of six varieties. One variety is rough and free from hair-like lines, has a quicksilver-like fracture and breaks when bent. Another variety breaks with difficulty and presents a sharp edge. The details of other varieties do not seem to be mentioned in the text. The following has been written on *kānta*:

> There are five kinds of it, namely *bhrāmaka, chumbaka, karṣaka, drāvaka* and *romakānta*. It possesses one, two, three, four and five faces and often many faces (with which to attract iron) and is of yellow, black and red colour respectively. The variety which makes all kinds of iron move about is called *bhrāmaka*; that which kisses iron is called *chumbaka*, that which attracts iron is called *karṣaka*, that which at once melts the iron is called *drāvaka* (literally a solvent); and the fifth kind is that which, when broken, shoots forth hair-like filaments.[141]

One is not quite sure if *kānta* can be considered proper iron metal. This kind of literature, however, suggests a classificatory system of iron in early India. The evidence of *Rasaratnasamuchchaya* is admittedly late but may contain an earlier tradition.

In a presumably late corpus entitled *Rasajālanidhi or Ocean of Indian Chemistry and Alchemy*[142] metals are said to be of seven kinds—gold, silver, copper, iron, zinc, tin and lead. The mixed metals are said to be three in number, viz., brass, bell-metal and *vartaka (suvarṃ a tārarkalauhaṃcha yaśodatrapusīsakam saptaitāni cha lohaṃ śuddhāni kṣitimanḍale/tritayaṃ miśralohaṃcha pittalaṃ kāṃsyavarttakam)*.

The Early Buddhist Canonical Literature in Pali

The previous section has discussed exclusively the evidence of the Sanskrit literature on iron. The chronological problems of the

[141] Ibid: 182.
[142] B. Mukherji, 1926 (?), II: 237.

individual texts were not generally mentioned in this discussion, which was perhaps not strictly necessary in our context. The chronology which is by and large accepted for the corpus of the Vedic literature puts it between c . 1500/1200 B.C. and c. 700/600 B.C. Pāṇini falls in the fifth century B.C. and Patanjali belongs to the fourth century B.C. or later. The *Arthaśāstra* may be placed in the Mauryan period. The epics were compiled between the late centuries B.C. and the early centuries A.D. This no doubt is a highly generalized chronological scheme, but from the present point of view—which is only to discuss the nature of the literary sources on iron—this generalized scheme may serve the purpose.

The chronology, of course, is the sore point even in the context of the early Buddhist canonical literature in Pali, but for the present purpose it should be enough to refer to Maurice Winternitz: 'at some period prior to the second century B.C., probably as early as the time of Aśoka or a little later, there was a Buddhist Canon which, if not entirely identical with our Pali canon, resembled it very closely.[143] Taking this considered opinion as the starting point one can aruge that it is improbable that the entire canonical literature was laid down only during this period and that there was no material going back to the sixth century B.C. i.e., to the time of the Buddha himself. The problem, however, is to determine how much of the material is immediately related to the Buddha or to the time immediately thereafter and how much of it comes down to the period marked out by Winternitz. One is not aware if any definitive literary and chronological stratigraphy of the early Buddhist literature has been established, but taken as a whole, there is no reason why the early Buddhist canonical literature or at least, the bulk of it, cannot be utilized as a source of the cultural material of India between 600 and 200 B.C. Among the recent writers who are by no means supporters of a high chronology, K. R. Norman[144] has argued that the *Therī-gāthā*, a collection of charming lyrics of the Buddhist female elders, was composed between the end of the sixth and the middle of the third centuries B.C. Even those who are wary of accepting anything in the *Jātakas* as datable cultural material may take note of the fact that quite a few of the *Jātakas* figure in clearly identifiable forms in the Indian art of the second century B.C. Unless otherwise stated, the

[143] Winternitz, 1933: 18.
XXIX.
[144] Norman, 1969: XXVIII–XXIX.

following references to iron in the Buddhist canonical literature may be taken as belonging to this broad time-span of c. 600–200 B.C.

The Anguttaranikāya contains the following references to iron:
1. 'Just as from an iron slab, heated and beaten all day, a bit may come off and cool down . . .' (*divasasantatte ayokapāle hannamāne . . .*).[145]
2. 'Place the body of the ranee, Bhaddā, in an oil vessel made of iron and cover it over with another iron vessel . . .' (*Bhaddāya deviyā sariram āyasāya teladoṇiyā pakkhiptvā annissā āyasāya doṇiyā paṭikujji*).[146]
3. 'Red-hot iron spike' (*ayosaṇku*), 'red-hot iron couch or bed' (*ayopaṭṭa*).[147]
4. Iron, copper, tin, lead and silver (*ayo, loham, tipu, sīsam, sajjham*).[148]

The relevant references in the *Majjhimanikāya* are the following:
1. 'As there comes to be an exceedingly loud noise from the roaring of a smith's bellows' (*seyyathā pi nāma kammāragaggariyā dhamamānāya adhimatto saddo hoti*).[149]
2. 'As . . . a strong man might cleave one's head with a sharp-edged sword' (*balavā puriso tiṇhena sikharena muddhānaṃ abhimanthayya . . .*).[150]
3. 'Bronze bowl, brought back from a shop or smithy covered with dust and dirt' (*kaṃsapāti ābhatā āpaṇā vā kammārakulā vā rajena ca malena*).[151]
4. 'Just like an iron hook stuck in a man's throat' (*ayosiṇghāṭakaṃ kaṇṭhe vilaggaṃ*).[152]
5. 'Hot iron stake' (*tattaṃ ayokhīlaṃ*).[153]
6. 'Encircled by an iron wall, with a roof of iron above, its incandescent floor is made of glowing iron' (*ayopākārapariyantoayasā paṭikujjito tassa ayomayā bhūmi jalitā tejasā yutā*).[154]

In the *Saṃyuttanikāya* there are only three references:
1. an iron pot, heated all day long (*divasaṃ santatte ayokaṭāhe*);[155]

[145] Woodward and Hare, 1932–6, IV: 41; Morris and Hardy, 1885–1900, IV: 70.
[146] Woodward and Hare, 1932–6, IV: 48; Morris and Hardy, 1885–1900, III.
[147] Woodward and Hare, 1932–6, IV: 87–8; Morris and Hardy, 1885–1900, IV: 130.
[148] Woodward and Hare, 1932–6, III; 11; Morris and Hardy, 1885–1900, III: 16.
[149] Horner, 1954–9, I: 297; Trenckner, 1888: 243.
[150] Horner, 1954–9, I: 298; Trenckner, 1888: 243.
[151] Horner, 1954–9, I: 32; Trenckner, 1888: 25.
[152] Horner, 1954–9, II: 61; Trenckner, 1888: 393.
[153] Horner, 1954–9, III: 227; Chalmers, 1899: 183.
[154] Ibid.
[155] Woodward, 1927, IV: 123; Feer, 1894: 190.

2. iron pin and spike (*ayosalākayā* and *ayosaṃkunā*);[156] and
3. 'Just as, Ānanda, an iron ball, if heated all day long, is lighter and softer, more plastic and more radiant...' (*sayyathāpi Ānanda ayogula divasam santatto lahutaro ceva hoti mudutaro ca kammaṃyataro ca pabhassarataro ca*...).[157]

The *Suttanipāta* refers to such items as iron goad (*ayos aṃkusa*),[158] stake (*ayosūla*),[159] ball (*ayogula*),[160] and hammer (*ayokuṭehi*).[161] In one case, however, *lohamayaṃ pana kumbhim* has been translated as an iron pot.[162] A very important reference is the following: '...for as a ploughshare that has got hot during the day when thrown into the water splashes, hisses and smokes in volumes, even so...' (*seyyathā pi nāma phālo divasasantatto udake pakkhitto cicciṭayati ciṭiciṭāyati sandhūpayati sampadhūpayati, evam eva*..).[163]

The above reference is important because it clearly shows that by the time of the *Suttanipāta* a metal poughshare could be the subject of an easy analogy, and considering the context, the metal concerned was in all probability iron. D. D. Kosambi[164] pointed out that the *Suttanipāta* or at least the part in which this reference occurs was one of the oldest parts of the Buddhist literature and thus it went back to 600 B.C. It is interesting to note that this reference occurs in exactly the same form also in the *Mahavagga* VI.26.7 of the *Vinaya* texts.[165] There is a reference to armour in *Mahavagga* X.2.4 (*vammikā*).[166] There are references to sword, shield, bow and quiver in *Cullavagga* VII.3.7: *asicammaṃ gahetvā dhanukalāpaṃ sannayithitvā*.[167] In *Cullavagga* VI.15.2[168] and V.28.1[169] there are references to *lohakaṭāha, lohakumbhī, lohabhāṇḍa*, etc., but as it is *loha*, in these cases one is not quite sure if they should be iron cauldron, pitcher, pot, etc.

[156] Woodward, 1927, IV: 206; Feer, 1894: 170.
[157] Woodward, 1930, V: 253; Feer, 1894: 282–3.
[158] Andersen and Smith, 1913: 128–9; Fausboll, 1881: 123.
[159] Ibid.
[160] Ibid.
[161] Ibid.
[162] Ibid.
[163] Andersen and Smith, 1913: 15; Fausboll, 1881: 14.
[164] Kosambi, 1963.
[165] Oldenberg, 1879: 225; Rhys Davids and Oldenberg, 1881–5, XVII: 95.
[166] Oldenberg, 1879: 342; Rhys Davids and Oldenberg, 1881–5, XVII: 295.
[167] Oldenberg, 1880: 143; Rhys Davids and Oldenberg, 1881–5, XX: 155.
[168] Oldenberg, 1880: 170.
[169] Ibid: 135.

The *Dhammapada* mentions *ayogula*[170] but more interesting is: 'as rust sprung from iron eats into its own source' (*ayasā va malan samuṭṭhitan taduṭṭāya tam eva khadati*).[171] *Ayogula* or iron ball occurs in the *Therīgāthā*[172] and a smith's daughter, Subhā, is mentioned,[173] but a hook (*aṅkusam*)[174] and a stake (*sūla*)[175] are mentioned as well. The *Theragāthā* refers to sickles, ploughs, curved spades (*asitāsu mayā naṅgalāsu mayā khuddakuddālāsu mayā*)[176] and razor (*khuram*).[177] There is also a reference to armour[178] ('I myself am binding on my armour'—*esa bandhāṃ sannāham*). *Ayogula* is all that one can get in the *Iti-vuttaka*[179] while both *Pettavatthu*[180] and *Niddesa II Cullaniddesa*[181] repeat the description of the *Niraya* hell noted earlier in the context of the *Majjhimanikāya*: 'encircled by an iron wall, with a roof of iron above, its incandescent floor is made of glowing iron.' The *Udānam* refers to a blazing spark of fire, struck from the anvil (*ayoghana*).[182]

The *Jātakas* also do not tell much. In the *Ghatajātaka*[183] there are references to *ayanangalāṃ* (iron ploughs), *ayakhāṇuke* (iron posts) and *ayasaṃkhalikaṃ* (iron chains). The context as a whole suggests a preoccupation with the magical properties of iron: 'if . . . four of you bring great iron ploughs and at the four gates of the city dig great iron posts into the ground, and when the city begins to rise, if you will fix on the post a chain of iron fastened to the plough, the city will not be able to rise.' There is a separate *Jātaka* called the *Ayoghara Jātaka*.[184] In the *Abbhantara Jātaka*[185] a tree is encircled with seven iron nets; the text uses the word *lohajāleni*. *Loha* may mean iron in this context. In the *Jarudupāna Jātaka*,[186] however, *aya* is iron while

170 Thera, 1914: 14; Max Muller, 1881: 74.
171 Thera, 1914: 35; Max Muller, 1881: 60–1.
172 Norman, 1971: 48; Oldenberg and Pischell, 1966: 247.
173 Norman, 1971: 35; Oldenberg and Pischell, 1966: 158.
174 Norman, 1971: 8; Oldenberg and Pischell, 1966: 128.
175 Norman, 1971: 9; Oldenberg and Pischell, 1966: 129.
176 Norman, 1969: 6; Oldenberg and Pischell, 1966: 7.
177 Norman, 1969: 72; Oldenberg and Pischell, 1966: 73.
178 Norman, 1969: 55; Oldenberg and Pischell, 1966: 57.
179 Woodward, 1948: 180; Windisch, 1889: 90.
180 Minayeff, 1888: 9.
181 Stede, 1918: 169, 170.
182 Steinthal, 1885: 93.
183 Fausboll, 1877–96, IV: 83; Cowell, 1895–1907, IV: 53.
184 Fausboll, 1877–96, IV: 491; Cowell, 1895–1907, IV: 305.
185 Fausboll, 1877–96, II: 397; Cowell, 1895–1907, II: 271.
186 Fausboll, 1877–96, II: 295; Cowell, 1895–1907, II: 205.

loha is copper. In the description of the hell meant for a parricide, the *Saṃkicca Jātaka*[187] refers to dome of iron, iron wall, iron cauldron, iron ball, iron ploughshare, iron beak, iron jaw, iron mouth, etc. Iron cauldron also figures in the *Nimi Jātaka*.[188] The *Dhamma Jātaka*[189] says, 'by iron gold is beaten'. The word is *loha*. The reference in the *Chhadanta Jātaka*[190] is possibly the most important of all these *Jātaka* references. 'She summoned smiths and gave them an order and said, "Sirs, we have a need of an axe, a spade, an auger, a hammer, an instrument for cutting bamboos, a grasscutter, an iron staff, a peg, an iron three-pronged fork . . .".' The set of implements is called to be of *asiloha*. In fact, this seems to be a completely new term for iron.

Among the non-canonical texts, the *Milindapanho* of the first century B.C. deserves attention. The ubiquitous 'red-hot ball of iron' (*tattaṃ ayogulaṃ*) occurs twice.[191] There is a reference to 'iron ore, and copper and brass and bronze' (*kālalohaṃ, tambalohaṃ vaṭṭalohaṃ, kaṃsalohaṃ*).[192] There is another reference to 'workers in gold and silver and lead and tin and copper, and brass and iron' (*suvaṇṇakārā, sajjhakārā, sīsakārā, tipukārā, lohakārā, vaṭṭakārā, ayakārā*). The translator, T. W. Rhys Davids, translates *vaṭṭa* as brass but adds, 'I can only guess what it is.'[193] One passage says, '*yathā mahārāja kālāyaso sudhito va vahati*' (O king, as black iron even when beaten out, carries weight).[194] In the same passage is added, '*kālayaso sakiṃ pītaṃ udakaṃ na vamati*' (as black iron does not vomit up the water it has once soaked in).

The Jain Canons

A survey of the cultural material in the Jain canons has been made by J.C. Jain.[195] He points out that the canons collectively do not belong to one particular period and that each part of the canon should be judged on its own merit after going through its contents carefully. The canons were written down by Devardhagaṇi in the sixth century

187 Fausboll, 1877–96, V: 266–370; Cowell, 1895–1907, V: 137–9.

188 Cowell, 1895–1907, VI: 59.

189 Cowell, 1895–1907, IV: 65; Fausboll, 1877–96, IV: 102.

190 Fausboll, 1877–96, V: 45; Cowell, 1895–1907, V: 25.

191 For II. 2.5, Rhys Davids, 1890: 70, Trenckner, 1880: 45; for III. 2.7, Rhys Davids, 1890 : 129, Trenckner, 1880 : 84.

192 Rhys Davids, 1894: 102; Trenckner, 1880: 267.

193 Rhys Davids, 1894; 102.

194 Rhys Davids, 1894: 364–5; Trenckner, 1880: 414–15.

195 Jain, 1947.

A.D. But it may be noted that the Mathura inscriptions of the reign of Kaniṣka clearly prove the existence of an organized Jain community whose organizational aspects can be matched in the canons. Some of the more important references to iron in the Jain canons may be cited following J. C. Jain.

'Iron was converted into steel and various tools and weapons—knives (*pippalaga*), needles (*sūi:āriya*), nail-cutters (*nakkhaccaṇi*) and surgical boxes (*satthakosa*), also coats of mail.'[196]

'Iron furnaces (*ayokoṭṭha*) are referred to which were filled with ore and a man handled it with tongs (*saṇḍasī*), then it was taken out and put on the anvil (*adhikaraṇī*).'[197]

'Iron was malleated, cut, torn, filed and was moulded by the blacksmiths.'[198]

The *Uttarādhyana sūtra* puts in the mouth of a prince called Mrigaputra some references to iron and iron implements. This prince was bent on becoming a *śramaṇa*.

'I have been forcibly yoked to a car of red-hot iron full of fuel . . .'[199]

'When I was born in hell for my sins, I was cut, pierced and hacked to pieces with swords and daggers, with darts and javelins.'[200]

The Evidence of the Graeco-Roman Literature

One may begin with the testimony provided by Ktesias about whose work J. W. McCrindle reminds us, 'it was the only systematic account of India the Greeks possessed till the time of the Makedonian invasion'.[201] His was the first special treatise on India for the Greeks who had no prior knowledge of the subject except for the report of Herodotus and the geographical account left by Hekataios of Miletos. A native of Knidia and belonging to a family practising medicine, he went to Persia in 416 B.C. and served eleven years under Darius II and six years under Artaxerxes Mnemon. He went back to Greece in 398 B.C. Among other things, he wrote a number of historical and geographical treatises on Persia. His work on India, like his other works, has been lost but has survived in the form of an abridgement done by Photios who was the Patriarch of Constantinople between A.D. 858 and A.D.886. Several fragments of Ktesias' work on India have survived in the works of other writers as well. To a Knidian

[196] Ibid: 100.
[198] Ibid.
[199] Jacobi, 1895: 94.
[200] Ibid.
[201] McCrindle, 1882: 4.

doctor in the Achaemenid court, India was, of course, a reality but it was never close enough to come into a distinct focus.

The following fragments refer to Indian iron.

Fragment I.4: reference to iron found at the bottom of a fountain (one which yielded gold also), adding that he had in his own possession two swords made from this iron, one given to him by the king of Persia, Artaxerxes Mnemon, and the other given by Parysatis, the mother of the same king. This iron, he says, if fixed in the earth, averts clouds and hail and thunderstorms; and he avers that he had himself twice seen the iron do this, the king on both occasions performing the experiment.[202]

Fragment I.22: the Kynokephaloi who live in the mountains . . . also sell swords, bows and javelins.[203]

Fragment I.27: for catching this worm in the river (it may be crocodile) a large hook is employed, to which a kid or lamb is fastened by chains of iron.[204]

Fragment 1.30: reference to a fountain where nothing sank except iron and silver and gold and copper.[205]

Among the historians accompanying Alexander, Quintus Curtius referred to the envoys from Malli and Sudracae in the following fashion.

> After the envoys of the Indians had been sent home, they returned a few days later with gifts. They consisted of 300 horsemen, 1030 chariots, each drawn by four horses abreast, a quantity of linen cloth, 100 Indic shields, 100 talents of white iron, lions and tigers of unusual size (previously tamed), also some skins of huge lizards, and shells of tortoises.[206]

Nearchus, the commander of the Greek fleet under Alexander, reported that the people of the Makran coast were not familiar with the use of iron. 'They carried thick spears about 6 cubits long, not headed with iron, but what was as good, hardened at the point by fire.'[207] And 'Things of a hard consistency they cut with sharp stones, for iron they had none.'[208]

In the context of the Mauryan India, Megasthenes wrote the following:

[202] Ibid: 8–9.
[203] Ibid: 24.
[204] Ibid: 28.
[205] Ibid: 31.
[206] Rolfe, 1946, II: 431–2.
[207] McCrindle, 1879: 183.
[208] Ibid: 184.

> It (India) contains much gold and silver, and copper and iron in no small quantity, and even tin and other metals, which are employed in making articles of use and ornament, as well as the implements and accoutrements of war.[209]
>
> The fourth caste consists of the Artisans. Of these some are armourers, while others make the implements which the husbandmen and others find useful in their different callings. This class is not only exempted from paying taxes, but even receives maintenance from the royal exchequer.[210]
>
> .... it is a practice with them to control their horses with bit and bridle ... they neither, however, gall their tongue by the use of spiked muzzles, nor torture the roof of their mouth.[211]

In addition, there is a specific reference in Megasthenes to the blacksmiths along with the woodcutters, carpenters and miners.[212]

In connection with the Indian war equipment, Arrian observed the following:

> They do not put saddles on their horses, nor do they curb them with bits like the bits in use among the Greeks or the Kelts, but they fit on round the extremity of the horse's mouth a circular piece of stitched raw ox-hide studded with pricks of iron or brass pointing inwards, but not very sharp; if a man is rich he uses pricks made of ivory.[213]

A major piece of reference occurs in the Periplus of the Erythraean Sea. The province of Ariaca which is identified by W. H. Schoff[214] with 'the northwest coast of India, especially around the Gulf of Cambay; the modern Cutch, Kathiawar and Gujarat' used to send, along with Indian cloth, Indian iron and steel to the eastern coast of Africa. The most celebrated reference was made by Pliny according to whom the best iron was that made by the Seres. We have already noted that according to Schoff this is Indian iron.[215]

At this point it may be noted that *ferrum candidum* of which the Malli and Sudracae sent 100 talents' weight to Alexander is, according to J. C. Rolfe, the translator of Quintus Curtius in the Loeb Classical Library, a mixed product: 'it denotes the presence of an alloy, but whether for increasing the beauty or strength of the iron is

[209] McCrindle, 1877: 31.
[210] Ibid: 42.
[211] Ibid: 89.
[212] Ibid: 86.
[213] Ibid: 221.
[214] Schoff, 1974: 70.
[215] Schoff, 1915.

uncertain.'[216] Schoff adds a note on 'Indian Iron and Steel' and the following excerpt from him is interesting:

> Marco Polo . . . Book I, chap. XVII, mentions iron and *ondanique* in the markets of Kerman. Yule interprets this as the *andanic* of Persian merchants visiting Venice, an especially fine steel for swords and mirrors, and derives it from *hundwaniy* 'Indian' steel. Kendrick suggests that the 'bright iron' of Ezekiel XXVII, 19, must have been the same. . . . *Ferrum indicum* also appears in the lists of dutiable articles under Marcus Aurelius and Commodus. Samasius notes a Greek chemical treatise 'On the tempering of Indian steel'. Edrisi says, 'The Hindus excel in the manufacture of iron. They have also workshops wherein are forged the most famous sabres in the world. It is impossible to find anything to surpass the edge that you get from Indian steel'.[217]

The whole position of Indian iron and steel in the Roman trade with India has been discussed by E. H. Warmington.

> Fine swords made of Indian steel had been famous since the time of Ctesias, and the Roman trade in Indian iron and steel was an important one. Since Pliny says that the finest of all iron was sent by the Seres with their tissues and skins the natural conclusion is that this metal was from the province of Shan-Si in China or at least from Ferghana. But, as Schoff has pointed out, the *Periplus* does not indicate the exportation of silk and steel together at the same marts, and we must take Seres as being the Cheras of the Malabar coast; 'Indian' is the epithet applied by the *Periplus* and by others, and by the Digestlist. Now the *Periplus* gives Indian iron with sword-blades at Adulis and other African ports, and the author knew it came from the interior of Ariace, yet did not see the metal at any Indian port; so that the Indians sent it in their own ships to the Axumites, who kept the secret of production, perhaps allowing the Romans to attribute the metal to remote China . . . Their excellent Parthian metal was perhaps really Indian. Eventually they learnt the secret of production, for Saumaise points out a special Greek treatise on the tempering of Indian steel. Chowstow may be right in supposing that the bulk of Roman importation consisted not of large quantities of ore, but objects made of iron and steel. The Romans worked it into fancy cutlery, as Clemens shews, and perhaps into armour at Damascus (whither Indian metal was sent) and at Irenopolis. The excellence of the steel would heighten the value of the untouched iron, but neither would come from distant China.[218]

[216] Rolfe, 1946.
[217] Schoff, 1974: 70–1.
[218] Warmington, 1928: 257–8.

II. DISCUSSION

In the previous section we have tried to enumerate the Indian literary data, primarily up to the early centuries A.D., as a source for a history of iron in India. In this connection we have also taken into account the relevant Classical references to Indian iron and steel. In this section we propose to evaluate these data with reference to three major issues: the beginning of the use of iron in agriculture, the terms used to denote iron, and the metallurgical details, if any.

The Beginning of the Use of Iron in Agriculture

It is not possible to deduce anything positive from the Ṛgvedic references to the term *ayas*, sometimes translated as 'iron'. The Ṛgvedic references to *ayas* occur in the following contexts: (i) 7 times (*ṚV* 1.1.52.8, 1.57.3, 1.80.1, 1.81.4, 1.121.9, 10.96.3-4, 10.113.5) as an attribute of the thunderbolt of Indra; (ii) twice as an attribute of Indra himself (1.56.3, 10.96.8), which has been taken to mean that Indra is clothed in *ayas* armour; (iii) 7 times (1.58.8, 2.20.8, 4.27.1, 7.3.7, 7.15.14, 7.95.1, 10.101.8[219]) as an attribute of forts; (iv) twice (9.1.2, 9.80.2) to denote the *ayas*-fashioned homes of Indra and Soma; (v) twice (5.62.7, 5.62.8) to denote the *ayas*-pillared chariots of Mitra and Varuṇa; (vi) twice (1.88.5, 10.87.2) to describe the teeth of Agni and Marut; (vii) twice (4.37.4, 6.71.4) to describe the jaws of a horse and the sun; (viii) once (1.116.5) to describe a thigh; (ix) once (1.163.9) to describe the feet of a sacrificial horse; (x) twice (6.3.5, 6.47.10) as an edge and a blade; and (xi) a number of times as the material of a beak (10.99.8), point (6.53.5), arrow (8.101.3), axe (10.53.9), vessel (5.30.15), and knives (8.29.3). None of these references is conclusive regarding the meaning of *ayas*. The term may mean either copper-bronze or iron or both. If the feet of *ayas* of the sacrificial horse in 1.163.9 suggest horseshoes, which are usually of iron, one may recall the description of Indra's thunderbolt of *ayas* in 10.96.4: 'golden-hued, gold-coloured, yellow in his arms'. The description of the chariot of Mitra-Varuṇa as *ayas*-pillared 'when the sun is setting' is also inconclusive because the colour of sunset is reddish and that may suggest any red-hot metal, not necessarily iron. The references to arrowheads, knives and axes are equally inconclusive as they may occur in both copper-bronze and iron. Similarly, no conclusion can be reached from 6.3.5 where Agni 'whets his splendour like the edge of *ayas*'.

[219] In his translation Griffith uses the description 'forts of iron'.

One should also mention allusions to metallurgical activities, though even in these cases there is no positive clue to the identification of *ayas*. There are two references to *karmāraḥ* (10.72.2, 9.112.2) which, in spite of possessing a marked similarity to *kāmār*, a word denoting blacksmith in some modern Indian languages (cf. Bengali, Hindi), cannot be assumed to mean a blacksmith in the Ṛgvedic context. The verb *dham* occurs four times (4.2.17, 5.9.5, 10.72.2, 10.82.3) and, though a metallurgical activity is implied in the contexts, one cannot take it to mean 'welding' which would suggest iron. A controversial passage is 9.112.2 where the medicinal plants, feathers of birds, and shining rocks are associated with a smith. The exact meaning of the passage is obscure but the suggestion of a metallurgical activity is clear enough. Any attempt to read anything more positive in this is suspect. Equally suspect are the attempts to infer the existence of iron on the basis of such words as *kṣura* ('razor', cf. 1.116.10, 10.28.9), *khādi* (its meaning is not specific but it is similar in name to that of the iron bangle a Hindu married woman is supposed to wear, cf. 6.16.40, 2.34.2, 10.38.1), and *svadhiti* ('axe', cf. 9.96.6, 3.8.11) which was used for felling trees.

It should be clear that any controversy regarding the meaning of *ayas* in the *Ṛgveda* or the problem of the Ṛgvedic familiarity or unfamiliarity with iron is pointless. There is no positive evidence either way. It can mean both copper-bronze and iron and, strictly on the basis of the contexts, there is no reason to choose between the two. Radomir Pleiner concludes after his survey of the Ṛgvedic evidence that '*ayas* . . . served as a term for a copper alloy or metal in a broader sense'.[220] While fully agreeing with the idea that *ayas* was a generic term, we believe that the term could include iron also.

The use of the terms 'black (metal)' and 'black *ayas*' in the *Black Yajurveda* clinches the issue. Purely on the basis of the literary data, iron may be considered a familiar metal at least in the *Doab* and the Indo-Gangetic divide, the basic geographical locale of the *YV* in c. 800 B.C. The exact point of the beginning remains uncertain. 'Black *ayas*' and 'dark metal' occur in the *Atharvaveda as well, but the AV* is likely to be a somewhat later text than the *YV* and thus the *AV* references are not of particular historical value.

Historically, the problem of the beginning of the use of iron in agriculture is a more significant issue than the problem of its very beginning. The crucial evidence in this regard, as we have already

[220] Pleiner, 1971: 9.

noted in the previous section, is that of the *Śatapatha Brāhmaṇa* (13-2.2.16-19 and 13-3.4.5) which associates iron with the peasantry and people in general. The date of thc *ŚB* is supposed to be c. 700 B.C. so that the association of iron with the common people and thus with agriculture in the Gangetic valley around 700 B.C. should, in fact, be beyond dispute. The *AV* reference (10.6.2-3) to an amulet made of metal taken away from a ploughshare also suggests that the ploughshare could be of iron, and thus the *ŚB* evidence is reinforced. The *AV* should be an earlier text than the *ŚB*. D. D. Kosambi put the *Kasibhāradvājasutta* of the *Suttaṇipāta* at around 500 B.C. and the clear reference to iron ploughshare in this *sutta* may be taken to represent conditions in the middle Gangetic valley in this period, if not earlier.

Terminology

1. The meaning of *ayas* in the *ṚV* is uncertain, but it is a generic term meaning possibly both copper-bronze and iron. It is unlikely that it meant only iron.
2. In the *YV* there are three relevant terms: *loha* (cf. *TS* 4.7.5) or *lohitāyasam* (cf. *MS* 2.11.5) meaning specifically 'copper-bronze'; *śyāmam* (cf. *TS* 4.7.5, *MS* 2.11.5, 4.2.9) meaning specifically 'iron' and plain *ayas* (cf. *TS* 1.2.11.f, etc.) which may still be a generic term but surely means 'iron' in some cases.
3. *Śyāmam* in the specific sense of 'dark metal' or iron occurs twice in the *AV* (9.5.4, 11.3.7) and *ayas* which may still be a generic term (but meaning 'iron' in many cases) occurs a number of times (cf. 8.3.2). *Śyāmam* occurs also in the *Vājasaneyī Saṃhitā*.[221]
4. The *ŚB* seems to use *ayas* for 'iron' in all cases. *Loha* and *Lohāyas* mean copper-bronze (cf. *ŚB* 5-4.1.1, 13-2.2.16).
5. The *Chhāndogya Upaniṣad* 6.1.6 mentions *kārṣṇāyasam* or black *ayas* in the specific sense of iron, but the references to plain *ayas* in the *Maitrī* 6.27 and 3.3 suggest iron.
6. Pāṇini seems to mean 'iron' by *ayas* in all cases (cf. 5.2.76, 3.3.82, 5.4.94, 8.3.48, 8.3.46, 5.4.143) but he also takes *ayas* both in the sense of 'genus' and 'species' (5.4.94).
7. Patañjali uses *ayas* always in the sense of 'iron' (cf. 2.4.10, 4.1.1., 3.1.7).
8. The *Mbh* (cf. 7.28.4) refers to *kārṣṇāyasam* but the general

[221] Weber, 1852: 566–7.

meaning of *ayas* seems to be 'iron' (cf. 4.38.26). *Ayas* in the sense of iron is mentioned also in the *Rāmāyaṇa* (3.35.35).

9. *Ayas* means iron in the *Manusmṛti* (cf. 2.168).[222]

10. In the *Arthaśāstra kālāyasa* is mentioned at least thrice (2.17.14, 3.17.7-8, 2.13.55) but plain *ayas* also is used to denote iron (cf. 4.1.62, 2.19.10, 13.2.23). There are two more relevant terms: *tīkṣṇa* (2.12.15; the original is *tīkṣṇadhātu* which has been translated as 'iron ore') and *vṛtta* (2.17.14, 3.17.7-8, 4.1.35) which has been translated as 'steel' by Kangle but as 'bronze' by Shamasastry.

11. In the Buddhist canonical literature *ayo* by and large means 'iron'. *Asiloha* as a term for iron occurs in the *Chhadanta Jātaka*. As far as one is aware *ayo* as iron occurs also in the Jain canonical literature. The *Milindapañho* denotes iron by *ayo* but it also mentions *kālaloham* and *kālāyasa*. It also mentions *vaṭṭa* or *vṛtta* of the *Arthaśāstra*.

To sum up the evidence on iron terminology in early Indian literature, *ayas* was used as a generic term in the *ṚV*, *YV*, and the *AV*, and in the last two texts there is no doubt that it included iron in some cases. From the *ŚB* onwards it has meant only iron in almost all the texts. Iron denoted as black metal in forms like *śyāmam, śyāmāyasa, kārṣṇāyasa, kālaloha, kālāyasa*, etc. has been mentioned only sporadically. Two other terms are mentioned: *asiloha* and *tīkṣṇa*. *Tīkṣṇa* possibly meant 'steel'. The exact connotation of *asiloha* of the *Chhadanta Jātaka* seems to be uncertain. *Vaṭṭa* or *vṛtta* also does not seem to be a precisely defined term. Pleiner has pointed out that, in the Pali commentary of *Vibhaṅga Athakathā* which is not earlier than the fifth century A.D., the term *vaṭṭaloha* occurs under the category of mixed metals along with brass and bell-metal.[223] It is also interesting that in *Rasajālanidhi*, a much later compilation of chemical and alchemical literature, *vartaka*, a variant of *vṛtta*, is put in the category of mixed metals. What also deserves notice is that this terminological evidence on iron in the early literature seems to have continued in the later texts. For example, terms like *kālaloha, kālāyasa, tīkṣṇa* occur in the Prakrit text of *Aṅgavijjā* (said to have been compiled in the fourth century A.D.),[224] and the lexical literature of *Amarakoṣa* and *Abhidhānaratnamālā*.

[222] Javaji, 1909.
[223] Pleiner, 1971: 14, note 66.
[224] Punyavijayji, 1957: 233.

Metallurgical Details

In the whole range of early Indian literature there is not a single reference to any metallurgical process which can be specifically associated with iron. Quenching may be suggested by the *Kasibhāradwājasutta* of the *Suttanipāta*, which refers to the immersion of a heated ploughshare in water. This may also be suggested by the following passage in the *Milindapañho*: 'as black metal does not vomit up the water it has once soaked in.' But references like this are virtually of no significance. There is no doubt that by the early centuries A.D. India was making steel which was also exported to the Roman world. In fact, the Classical testimony which we have discussed is fairly consistent on this point. Indian iron was considered the best, and the Gujarat coast was the region through which Indian iron objects used to be exported to the east coast of Africa. But it needs to be emphasized that nowhere in early Indian literature is there a specific reference to steel-making. *Tīkṣṇa* seems to mean 'steel' and its earliest occurrence is in the *Arthaśāstra*. This however does not mean that steel came to be manufactured only about this period.

We have cited two passages from the *khaḍgalakṣaṇam* section of Varāhamihira's *Bṛhatsaṃhitā* on the treatment of iron blades. These read very much like some of the recipes in the late mediaeval European literature on the subject. Compare an excerpt from a sixteenth-century text on how to harden steel and make good cutting edges: 'take the leaves and the root of the plant called oxtongue, boil them in water, and quench therein.'[225]

Finally, it may be noted that there is no reference to what may be interpreted as cast iron in Indian literature. P. C. Ray suggested that *tīkṣṇa* could be translated as cast iron as well but could not cite any argument.

[225] Smith, 1968: 11.

CHAPTER FIVE

The Pre-industrial Iron-smelting Tradition

I. DISTRIBUTION

The basic sources of this study are different nineteenth-century records referring to pre-industrial iron-smelting—travellers' records, Geological Survey of India *Memoirs* and *Records*, and occasional metallurgical publications. Many of these discussions later on found a place in various *District Gazetteers*. Perhaps the most convenient introduction to the theme is the section on iron in Valentine Ball's *Manual of the Economic Geology of India* published in 1881. Ball's intention was quite clear: 'As will be abundantly shown in the course of the following pages, the manufacture of iron has in many parts of India been wholly crushed out of existence by competition with English iron, while in others it is steadily decreasing and it seems destined ultimately to become extinct. For this reason alone, if there were no others, the native process is worthy of full notice here.'[1]

The distribution of smelting centres is obviously related to the distribution of suitable ores. We have already noted that the pre-industrial iron-smelters could get iron not merely from the large and well-known deposits but also from ferruginous laterite and quartz-iron-ore schist. We have also taken cognizance of the fact that the ores suitable for pre-industrial iron-smelting occur all over the Indian subcontinent outside the major alluvial stretches. This distribution includes Sind, Baluchistan, the Northwest Frontier zone, Kashmir, the Salt Range and Kot Kerana hills of Panjab, the Uttar Pradesh

[1] Ball 1881: 338.

and Panjab Himalayas, Patiala, Rajasthan, Gujarat, central India, Maharashtra, Mysore, Kerala, Tamil Nadu, Andhra Pradesh, Orissa, West Bengal, Assam, Bihar and Mirzapur in Uttar Pradesh.

According to Ball's list, which is doubtless the most comprehensive list available, pre-industrial iron-smelting has been reported from the following districts, regions and former princely territories: Travancore state, Madura, Thiruchirapalli, Coimbatore, Salem, North and South Arcot, the Malabar and Nilgiri regions, Chingleput, Tumkur, Mysore proper, Shimoga, Kadur, Chitaldurg, Cuddapah, Kurnul, Bellary, Krishna and Godavary districts, Hyderabad state, Vizagapatnam, Balasore, Talcher, Mayurbhanj, Birbhum, Bankura, Burdwan, Purulia, Monghyr, Singhbhum, Hazaribagh, Lohardaga, Palamau, Ranchi, Dhanbad, Sambalpur, Bilaspur, Raipur, Mandla, Bhandara, Balaghat, Jabalpur, Narsinghpur, Chanda, Mirzapur, Rewa, the Bundelkhand region, Banda, Lalitpur, Panna, Gwalior, Alwar, Jaipur, Ajmer, Udaipur, Nimar and Malwa, Ratnagiri, Satara, Surat, Kolhapur, Sawantwari state, Panchmahal, Kaira, Rewa Kantha, Ahmedabad, Kathiawar, Kutch, Sind, Bannu, Peshawar, Salt Range, Kot Kerana, Kangra, Mandi, Kulu, Sirmur, Gurgaon, Kumaun-Garhwal, Assam and the Khasi and Jaintia hills.

F. R. Mallet, a geologist of the Geological Survey of India (like Valentine Ball), contributed the section on 'Iron Ores of India' to George Watt's *A Dictionary of the Economic Products of India* (volume IV) which was published in 1890. In this section he also took note, briefly but pointedly, of the pre-industrial iron-smelting tradition in all the relevant localities of the subcontinent. The following review of the evidence is based entirely on his testimony.

Travancore: despite the presence of black magnetite sand in enormous quantities along the coast of south Travancore, the pre-industrial smelters of the region used chiefly laterite. 'The outturn, however, was not very great even more than forty years ago, the taluk of Shenkotta being the only one in which any considerable quantity was produced.'[2]

Tinnevelli: the area contains lateritic ore and magnetite. 'The ore that has been used in the native furnaces is magnetic ironsand, from which also steel has been produced in the Village of Vanga-colum.'[3]

Madura: there is an abundance of lateritic ores with considerable traces of earlier smelting, 'but the industry is now quite extinct'.

[2] Mallet, 1890: 505.

[3] Ibid.

Pudukottai: magnetite is known to occur in this area but Mallet does not specifically refer to pre-industrial smelting.
Trichinopali: ferruginous nodules were utilized for pre-industrial smelting at some former period, 'but no iron is made in the district now-a-days'.
Coimbatore: black magnetite sand was used by the pre-industrial smelters but in Mallet's time the tradition was almost, if not entirely, extinct.
Nilgiri Hills: haematite and magnetite are tolerably abundant but there is no specific reference to pre-industrial smelting by Mallet. He adds: 'the comparative scarcity of fuel renders it improbable that these ores could be utilized.'
Malabar: magnetite and laterite are the main ores. 'Iron has been produced on a considerable scale, the ore being obtained, and also smelted, in the taluks of Kurmenaad, Shernaad, Walluwanaad, Ernaad, and Temelpuram. In 1854 there were over 100 furnaces in operation, which were of a larger size than those commonly used in India, some being as much as ten feet in height. Nearly a ton of ore is said to have been the charge, from which, in round numbers, from 250 to 400 lb. of metal was obtained. At the date mentioned, the quantity of iron produced yearly in the district was estimated at about 475 tons, but these figures seem very small, considering the number of furnaces in blast.'[4]
South Kanara: iron ore is abundant in the region but Mallet does not specifically refer to pre-industrial smelting.
Salem: ores are said to exist in inexhaustible quantities. 'Excellent iron is produced in the Salem district, but, as in so many other parts of India, the outturn is decreasing, partly from the growing scarcity of charcoal, and partly from the influx of English iron.'[5]
South Arcot: there was smelting 'in a very large number of villages, more especially in the taluks of Trinomalai and Kallakurchi. The manufacture of steel has also been carried on'.[6]
North Arcot: 'black sand' and magnetite were the principal ores used. Around 1855 smelting operations were conducted in 86 villages, with ores being mined in every taluk.
Chingleput: magnetite occurred in the local hills.

[4] Ibid.
[5] Ibid: 506.
[6] Ibid.

Nellore: magnetite is said to be abundant but there are notices of haematite schist. In Mallet's time the smelting operations were extinct.
Kadapah and Kurnool: there is an abundance of ores. Haematite was mined extensively.
Anantapur: iron is said to be abundant.
Bellary: 'the supply of splendid haematite ore is absolutely unlimited' near the town of Bellary. 'The old iron industry is, however, nearly extinct, owing mainly to the scarcity of fuel.'[7]
Kistna: the area comprised the old collectorates of Guntur and Masulipatam and 'the principal localities are in the Guntur taluks of Datchapali, Timmarkatta, and Narsaranpett, and the taluks of Jaggiapetta, Tirwar, Gallapali, and Pentapand in Masulipatam'.[8]
Godavari: limonite and haematite occur in large quantity 'but it is only worked in a few localities, and on a small scale'.
Vizagapatam: iron has been manufactured in numerous places, the 'zamindari of Jaipur' being one of the principal seats of the industry.
Ganjam: although iron ores are known in this area, there was no smelting operation in Mallet's time.
Ashtagram division (Mysore): although abundant, iron ores are not mined in many places in this section.
Bangalore division: 'both iron and steel are produced, the ore principally used being "black sand". Chinapatam has long been celebrated for the manufacture of steel wire, which has been sent to remote parts of India for the strings of musical instruments'.[9]
Nagar division (Mysore): 'black sand' was the principal ore used for the local manufacture of iron and steel. In this context Mallet adds that 'in Mysore, as in so many other parts of India, the outturn of iron has greatly diminished during the last few decades'.
Hyderabad: haematite and magnetic iron are the main ore-types.

> Smelting has been carried on extensively, in native furnaces, in the parganas of Kallur and Anantagiri. Haematite, titaniferous ironsand, and yellow and red ochre are stated to occur in *Warangul*. Steel-making has been carried on at several villages in *Yelgandal*, but one of the most celebrated places for its manufacture is Konasamundram in *Indor*, 12 miles south of the Godavari and 25 from the town of Nirmal. Accounts written some fifty years ago state that most of the outturn was bought by dealers from Persia who travelled to Konasamundram for the purpose,

[7] Ibid.
[8] Ibid.
[9] Ibid: 506–7.

> and that the steel was used in making the celebrated Damascus sword blades. Two kinds of iron were used in its production—one from Mitpalli smelted from 'ironsand', or from Dimdurti, reduced from magnetite occurring disseminate through gneiss and mica schist, and the other from Kondapur smelted from an ore 'found amongst the iron clay' (laterite?): three parts of the former were employed to two of the latter.[10]

Bastar: ores are said to be abundant and it is interesting that the deposit 'in immense quantities, on the Bela Dila' was already known in Mallet's time. The smelting operations were on a minor scale.
Bilaspur: the ores of the area were not worked on a major scale.
Raipur: the ores principally used are haematite and laterite. A considerable number of furnaces were at work in the Daundi-Lohara zamindari and the feudatory state of Khairagarh. The other areas which are mentioned are the zamindaris of Gandai, Thakurtola, and Worarband, and the state of Nandgaon.
Chanda: this has been said to be one of the most remarkable districts of India as far as the accumulation of iron ores (principally haematite, but magnetite and limonite as well) is concerned. An important locality is Lohara 'where haematite of extreme purity forms an entire hill ⅜ of a mile long, 200 yards broad, and 120 feet high'. Haematite was the material chiefly used by the pre-industrial smelters of the area.
Balaghat: pre-industrial smelting was very near extinction when Mallet wrote, although a few furnaces still struggled for existence in the Baila, Kini and Bhanpur zamindaris, using mainly haematitic rocks.
Bhandara: about the middle of the nineteenth century, 160 furnaces were at work using lateritic ores. The metal produced from the mines of Agri and Ambaghari in the parganah of Chandpur was said to be of excellent quality.
Nagpur: ore is said to occur but there is no reference to smelting.
Mandla: lateritic ores were used by the pre-industrial smelters of the district.
Seoni: pre-industrial operations were current in the Juni and Katangi areas.
Chhindwara: lateritic ore was known to exist but there is no specific reference to pre-industrial smelting by Mallet.
Nimar: some ores were known to exist but there is no reference to pre-industrial smelting.

[10] Ibid: 507.

Hoshangabad: haematite was worked in the area to the north-west of Harda and south of the Narmada.
Narsinghpur: the major iron-producing place in the district was Tendukhera and the following observation by Mallet is important.

> Tendukhera has long attracted notice owing to the excellent quality of the iron manufactured there, which is said to command a higher price than any other iron made in the Narmada valley. In 1830 a suspension bridge over the Bias river, in Sagar, was opened, the iron for which had all been smelted in native furnaces at Tendukhera. The ore is an earthy manganiferous haematite and limonite, and the good quality of the iron has been attributed to the ore being somewhat calcareous (the gangue being partly limestone), which produces the same effect as would a purposely added flux. The quality may also be due, in part, to the manganiferous character of the ore. It appears that part of the iron that is produced is converted into steel, by a method different from that practised in Madras.[11]

Jabalpur: in Mallet's time iron was 'still smelted on a somewhat considerable scale in the district'. Different varieties of ores occurred in immense quantities, with the Janli mine being the most extensively worked in the region. 'At Mangeli, Mogala, Gogra, and Danwai in the Lora ridge, a manganiferous haematite is worked, the iron produced therefrom being a hard steely kind used for edged tools. The iron smelted from the Agaria, Partabpur, and Janli is much softer. . . . The laterite ore used to be smelted on a considerable scale, but of late years the mines have been abandoned.'[12]
Rewa: ores are known but there is no reference to pre-industrial smelting.
Bundelkhand: rich haematite of the region was worked on a considerable scale, Herapur being one of the chief centres of production.
Gwalior: a number of iron mines occurred in the vicinity of Gwalior. 'About 50 miles to the northwest of Gwalior there is a forest, which it is estimated would, without replanting, supply fuel for an outturn of 12 tons of bar iron a day for a period of 900 years. The mines of Bagh (about 60 miles west-south-west of Indore) have long been celebrated.'[13]
Indore, Dhar, and Chandgarh: rich deposits of haematite occur but there is no reference to pre-industrial smelting.

[11] Ibid: 508.
[12] Ibid: 508–9.
[13] Ibid: 509.

Ali Rajpur: ores occur but there is no reference to pre-industrial smelting.
North Kanara: lateritic ore occurs in the Western Ghats but Mallet does not refer to pre-industrial smelting.
Dharwar: when fuel was plentiful much iron used to be smelted in the Kappatgudd hills.
Kaladgi: the principal ore is haematite but there is no reference to pre-industrial smelting.
Belgaum: the tradition of iron manufacture was extinct when Mallet wrote.
Goa: ores occur but there is no reference to pre-industrial smelting.
Sawantwari and Kolhapur: iron used to be manufactured from lateritic ore.
Ratnagiri: lateritic ore of the district was formerly smelted.
Satara: lateritic ore was utilized on a small scale.
Surat: ores occur but there is no reference to pre-industrial smelting.
Rewa Kantha: along the western limits of the district iron was extensively worked. Large heaps of slag still remained near Jambughoda, Limodra and Ladkesar.
Panch Mahals: ore of considerable richness is said to occur but there is no reference to pre-industrial smelting.
Kaira and Ahmedabad: heaps of slag suggest former smelting but in Mallet's time iron was not worked in these areas.
Kathiawar: Mallet wrote that even fifty years ago the manufacture of iron was in a moribund condition and that in his time there was not a single furnace in blast.
Cutch: in this area also the pre-industrial smelting tradition was extinct.
Sind: ores occur but there is no reference to pre-industrial smelting.
Marwar, Ajmere, Bundi, Kota and Bharatpur: ores occur but there is no specific reference to the tradition of pre-industrial smelting.
Mewar: limonite was worked to some extent in Mallet's time.
Jaipur: large quantities of ore exist but the workings were abandoned in Mallet's time.
Alwar: the local ore is said to be a mixture of limonite, magnetite and manganese oxide. The deposits are obviously rich, and Mallet referred to 'the numerous furnaces in the Alwar territory'.
Bannu: according to Mallet, the hills 25 or 30 miles south-east of Bannu contain haematite in abundance. The iron produced from it was in great demand at Kalabagh.

Peshawar: the magnetic ironsand collected from Bajaur used to be smelted in the region of Peshawar.
Jhelam: haematite is said to be abundant in the Kot Kerana hills but there is no direct reference to pre-industrial smelting.
Kangra: magnetite, magnetic ironsand and haematite are found in Kangra which used to possess a rich tradition of pre-industrial smelting. We here choose the observation on the Kangra and Mandi iron industry from the gazetteer of Kangra district in 1883–4:

> The ore . . . is of the same nature as the products of the best mines of Sweden, and it is worked, as there, at its outcrop in open quarries. It is one of the most valuable ores of iron, being readily reduced, in contact with charcoal, in furnaces of the simplest construction, and yielding the very best quality of iron. Some of the metal . . . was sent to England in 1858 for obtaining an estimate of its value . . . While best English iron yielded at a pressure of about 56000 1bs. on the square inch, the Kangra iron in the state in which it was sent . . . required a force of 61300 per square inch to break it.[14]

Mandi: it is basically the Kangra iron industry.
Simla hill states: magnetic ore and magnetic ironsand were chiefly used for smelting in a number of areas such as Bashahr, Jabbal, Dhami, Sirmur, Nahan, etc.
Gurgaon: Firozepur in the extreme south of this district once possessed considerable smelting works using chiefly haematite. Even in Mallet's time some iron was produced here.
Kashmir: Mallet wrote the following on Kashmir and Ladakh:

> Limonite has been extensively mined on the Punch river, in the outer hills. Iron is also worked in the neighbourhood of the Dragar mountain to the north of Pansir, in the Riasi district, at the village of Soap, or Sufahan, situated on the Bimwar river, at the southeastern end of the Kashmir valley, at Arwan, near Sopur, and Shahr near Pampur. In Ladakh, at the village of Wanla, nearly due south of Khalsi on the Indus there are very extensive iron works.[15]

Kumaun: rich and abundant deposits occur in several parts of Kumaun. 'Amongst the localities which have attracted most notice is Ramgarh, near which several mines (Pahli, Loshgiani, Natna Khan, Parwara, etc.) have been worked in micaceous haematite and limonite. Haematite and micaceous iron also exist near Khairna, but the deposits are not of much importance. Very large quantities which is

[14] Cited in Chakrabarti and Hasan, 1984.
[15] Mallet, 1890: 511.

earthy haematite and limonite, occur in beds in the Sivalik strata, in the neighbourhood of Kaladhungi and Dechauri. The Dechauri ore is of better average quality than the other.'[16]

Lalitpur: a considerable amount of iron is smelted from haematite at Salda in the Maraura parganah, while a hard steely kind is made at Pura.

Banda: there were mines at Deori and Khirani, and iron was extensively worked at several points in the Kalyanpur parganah, especially at Gobarhai.

Mirzapur: in Mallet's time the scale of smelting operations in Mirzapur was relatively minor, although magnetite was known in the southern part of the district.

Orissa: the districts or areas which have been mentioned by Mallet in this connection are Cuttack, Talcher, Dhenkanal and other tributary states.

Burdwan: iron ores of Burdwan (nodules of clay ironstone) have been mentioned by Mallet who, however, does not make any special reference to the pre-industrial smelting operations of the area.

Birbhum: Mallet's observation is as follows: 'Iron used to be made on rather a large scale, the *kacha*, or crude, metal unlike that produced in other parts of India, resembling pig-iron, from which *pakka*, or refined, iron was produced by a sort of puddling process. The ore used in the Birbhum Iron-Works Company's (a modern company established about 1855) furnaces was obtained from beds near the base of the laterite.'[17]

Bhagalpur and Monghyr: some smelting operations were conducted in both districts.

Gaya: smelting operations are mentioned by Mallet.

Hazaribagh: furnaces using the ironstone shale groups were numerous in the vicinity of the Bokaro and Karanpura coalfields but elsewhere in the same district other ores were used.

Manbhum: Mallet refers to 'the excellent quality of some of the native-made iron' of this district.

Singhbhum: the ores smelted in this district were chiefly from ferruginous schists, and from the laterite.

Lohardaga: a variety of ores occurs in this area (chiefly magnetite, haematite, limonite, and lateritic ores on the tops of the plateaus), but Mallet does not specifically refer to pre-industrial smelting.

[16] Ibid.
[17] Ibid: 512.

Tributary states of Chhotanagpur: according to Mallet iron-smelting was carried on to some extent.
Darjeeling: 'a valuable bed of extremely pure magnetite, with some micaceous haematite' was reported by Mallet five miles south-east of Kalimpong. 'The magnetite has been smelted on a trifling scale, and is said to yield a steely iron suitable for making *kukris* and *bans*.'[18]
Khasi-Jaintia Hills: 'Iron used to be made in the Khasi-Jaintia hills, but the manufacture has completely died out. The ore employed was a titaniferous magnetite, occurring in the form of minute grains disseminated through decomposed granite. The soft friable rock was raked into a narrow channel with a rapid current of water running down it, and the heavy ironsand caught by a small dam, while the lighter particles were washed away. After further purification by re-washing, the ore was smelted, and yielded an iron which, after refining more than once by reheating and hammering, was of excellent quality'.[19]
Naga Hills: Mallet refers to the smelting of clay ironstone in the tertiary coal-measures of the Naga hills, but he says that, although the outturn must once have been considerable it was all but extinct as early as 1841.
Manipur: a layer of clay containing small pisolitic nodules of limonite is the chief source of pre-industrial iron in Manipur. Titaniferous iron ore is obtained in some streams.

The foregoing inventory of the main places and areas of pre-industrial iron-smelting in India is important because it indicates what the general situation was like in about the middle of the nineteenth century and possibly some years later. If anything, Mallet has been cryptic in his references to the tradition of pre-industrial smelting and in some of the areas where he does not specifically mention the existence of this tradition (cf. Lohardaga) we know for certain that the tradition was very much there. The Lohardaga area, for instance, was the heartland of the Agaria and Asur iron-smelters till recently. But his inventory has the merit of offering a comprehensive subcontinental picture.

II. THE GENERAL METHODS OF SMELTING OPERATIONS

One of the most systematic discussions of the pre-industrial iron-

[18] Ibid.
[19] Ibid.

smelting processes in India may be found in John Percy's *Metallurgy: Iron and Steel*, published in 1864.[20] Percy isolates three basic types of pre-industrial iron-smelting furnaces.

The first was the simplest and also the most common. Circular in form, its height varied from two to four feet. At the bottom, or across the hearth, the width was from ten to fifteen inches, and at the top from six to twelve inches. It was made entirely of carefully tempered clay. The lower part tended to wear away rapidly and was constantly repaired with linings of fresh clay. There were two openings towards the bottom of the furnace—one for removing the cinder, and the other for drawing out the smelted product, or sponge-iron. Through the second opening two earthen pipes, or tuyeres, connectd with a pair of bellows, were inserted. Both the openings were covered with clay before the furnace was lit. The opening for the cinder was generally at the side, that for the sponge-iron and the tuyeres was in front. The tuyeres, some twelve inches long and an inch in internal diameter, were placed side by side, projecting two to three inches into the furnace, three to four inches from its bottom.

If the furnace was newly built, it was first dried by keeping a fire going in it for several hours. The tuyeres were then placed in the positions mentioned above, and the openings both at the side and front stopped with clay. The furnace was half-filled with charcoal and lighted and then filled up to the top. The bellows were applied at this stage. When the charcoal at the top had partly subsided, alternate charges of ore and charcoal were applied till the requisite amount of ore had been introduced. The blast from the bellows was then increased to the maximum and kept there till the operation was complete. This took four to six hours, during which the cinder was removed from time to time with the help of a small rod or bar through the opening devised for that purpose. But still the greater part of the cinder remained in the furnace and was removed with the sponge-iron, which was taken out at the end of the operation, taking off the front cover. If sufficiently hot, the sponge-iron was immediately hammered into a tolerably sound bloom, and if it came out too cold for the purpose, it was reheated and hammered.

The second type of furnace was a cavity made in a bank of clay which was well-tempered. The cavity was cylindrical, fifteen to eighteen inches in diameter, and some two and a half feet deep or high.

[20] Percy, 1864: 254–70.

At the bottom were two openings facing each other, through one of which the tuyeres were inserted. A row of such furnaces could be made at convenient distances from each other in a bank of clay—a distinct advantage the type offered. The furnace was filled with charcoal and lit in the manner previously described. Alternate charges of ore and charcoal were applied and the bellows kept going at full blast. When the cinder reached a certain height in the furnace, it was tapped with an iron bar inserted through the front opening. The smelted ore was first shaped into a ball with an iron bar introduced from above, then taken out with tongs through the top again. After the ball was removed the cinder was also completely removed by tapping through the front opening. The furnace was then ready for the next charge of ore and charcoal. The lower part of the furnace did not have to be (indeed, could not be) removed for a fresh charge, which saved time and was an improvement upon the first type of furnace. According to Percy this was 'in fact a small Catalan furnace'.

The third type of furnace mentioned by Percy was also a cavity scooped out in the side of a clay mound. Its height on the outside was eight to ten feet but inside the furnace was only six to seven feet high, the bottom being two to three feet above ground level. The internal diameter, top to bottom, was eighteen inches square, but a variation of fifteen by twenty-one inches was also known. The front wall was only five to six inches thick and could be removed at pleasure. When it was removed the furnace presented the appearance of a vertical trench cut in a mound of clay. The base of the furnace was provided with a perforated tile of dried clay placed at an angle of forty-five degrees to the back of the furnace.

The base tile or plate was first positioned and cowdung deposited up to a height of twelve inches—four or five inches above the upper edge of the plate. Above this bed of cowdung two earthern tuyeres, at least eighteen inches long, were introduced, almost touching the back of the furnace. The furnace was then partly filled with charcoal, lighted and then filled up to the top. The blast was applied from the front and the man working the bellows sat upon a sort of scaffold two to three feet from the ground. Ore and charcoal were then alternately introduced and the whole operation took twelve to sixteen hours. A considerable quantity of cinder was tapped at intervals during the operation with an iron bar passed through the perforations in the base plate, beginning with the lower holes and then proceeding to the upper ones. The holes through which the cinder had been drawn were stopped with clay as the iron accumulated at the bottom and would otherwise escape. When the iron rose to the level of the tuyeres and the tuyeres were burnt away, the smelting was considered complete. The base plate was then removed with an iron bar and the mass of cinder and iron allowed to fall to the ground. This lump of iron weighed a hundred and fifty to two hundred pounds, and was thus too large to be hammered whole. It was therefore cut by means of a sharp-edged sledge, so that when cold it could be broken into four pieces. It consisted of a mixture of malleable iron and natural steel, and their proportion depended on the nature of the ore. It has, however, been pointed out by Percy that, when the object was to produce steel, a large proportion of charcoal was employed and a gentle blast applied:

> The steely parts frequently present the same appearances on fracture as the best blister steel from Swedish iron; and they are carefully selected and prepared for use by being heated to a low red heat in a charcoal fire and then cut into small pieces of convenient size for making edge-tools, etc. When iron is wanted instead of steel, the pieces into which the lump has been broken are raised to a welding heat and hammered into bars, by which means it loses almost all appearance of steel. Sometimes small quantities of cast-iron are produced in this furnace to the great annoyance of the smelters who have much difficulty in separating it from the rest of the iron. They consider that in this case the iron has been injured by raising the temperature of the furnace too high.[21]

The third type of furnace was thus obviously used to make better-grade iron and natural steel.

A close study of records from different parts of the country reveals that there were regional variations of each of the three types of furnace described above. The variations are not important except in detail. However, in 1963, H. F. Cleere who was then Assistant Secretary of the Iron and Steel Institute, Britain, reported two types which do not seem to have been reported before. A demonstration of primitive Indian iron-making was staged at the National Metallurgical Laboratory, Jamshedpur, in February of that year, and Cleere saw three types of furnaces in operation then. This demonstration, held during a symposium on recent developments in iron and steel-making with special reference to Indian conditions, was arranged by Moni Ghosh, a retired employee of the Tata Iron and Steel Company,

[21] Ibid: 260; for the subsequent section on the Single Becha, Kamarjoda and Jiragora furnaces, Cleere, 1963.

who is said to have studied these primitive furnaces and their aboriginal operators for many years. These furnaces were called, after the three villages associated with them, the Kamar Joda furnace, the Single Becha furnace and the Jiragora furnace. The last conforms to the first type described by Percy.

The Kamar Joda furnace consisted of a shallow bowl, about a foot in diameter and six inches deep, excavated in the earth and lined with clay. A mound some eighteen inches high was erected above the bowl, extending two feet or so behind it. A shaft, three to four inches in diameter, was made immediately above the centre of the bowl, the front of which was open. The structure was entirely of puddled clay mixed with chopped straw or husk. It was dried in the sun and the cracks were lined with fresh clay. When the furnace was ready for operation, the bowl was filled with charcoal and the next step was to close the open front with damp sand, probably bonded with a little puddled clay. A tuyere was inserted through the sand-wall and the blast applied through twin foot-operated bellows. Some thirty pounds of ore were charged with an equal amount of charcoal. When the charge had been used up, the sand-wall was broken open and the bloom, a mass of slag and iron with much adhering unreduced ore and charcoal, dragged out with tongs and quenched with water. The bloom was then battered with a hammer and the small fragments of metal picked up by hand. These bits of iron were then reheated and forged together. Cleere points out that in this type of furnace temperatures did not exceed approximately 1100 degrees Centigrade and the metal was never molten during the process. As the ore was reduced to metal it became plastic and very slowly travelled down the furnace, forming globules of varying sizes. There was almost no accretion of carbon at the low temperatures encountered, and the charcoal used as fuel did not introduce sulphur or phosphorus in the metal. So the result was a very pure iron. An ingot of two pounds, approximately, was obtained from thirty pounds of ore. The Kamar Joda furnace has been described as a furnace of the most primitive type, because it had no provision for tapping of slag. The metal tended therefore to scatter in small discrete particles throughout the slag matrix, which had to be broken open to obtain the metal.

The first step in building the Single Becha type of furnace was to dig a pit, some four feet square and three feet deep with steps leading down to it. A bowl, a foot in diameter and nine inches deep was then dug in the floor of the pit, the bowl extending in part beyond one of the side-walls. The next step was to dig a shaft, three to four inches in diameter, immediately over the centre of the bowl. For tapping the molten slag a pit was dug out of the bowl, sloping downwards. This slag-pit lay inside the main pit. Hearth, shaft and slag-pit were lined with puddled clay mixed with chopped straw or husk and sun-dried, the cracks filled later with fresh clay. When the furnace was in operation, the operator probed the hearth at intervals through the tuyere with an iron rod. When the end of the rod showed slag the sand-wall blocking the slag pit was pulled down and the slag allowed to run out and collect in the pit, where it was quenched with water before being thrown out. This was done a number of times till the end of the operation. The spongy mass of iron was then taken out and after hammering off the adhering slag it was reheated or forged. The major advantage this furnace offered over the Kamar Joda furnace was that the tapping of the slag made for a greater coherency of the iron, and consequently less difficulty in manipulation for forging and shaping.

The Jiragora furnace shows a shaft which is about two and a half to three feet high with an internal section tapering from one foot at the base to three to four inches at the top. On the opposite side of the tuyere aperture a platform supported on bamboo poles is built on the top of the furnace. This is meant to carry the charge before it is fed gradually into the top of the furnace. Everything is built in clay. The front arch is closed with sand and the two bellows operating through a single tuyere supply the blast. Slag is removed at intervals by breaking open the sand front and the slag is allowed to collect in a pit nearby. Cleere describes this furnace as 'an above-ground version of the Single Becha furnace'.

It would now seem that more than the three basic types of furnaces (as postulated by Percy) were employed in primitive Indian iron-making. The parallels of the Kamar Joda furnace and the Single Becha furnace need further scrutiny on the basis of published records in different parts of India.

Certain general features may be observed at this stage. Ore, before being put in the furnace, was carefully sifted and reduced to the size of small peas. In some cases ore was roasted to increase its ferrous content. Charcoal of *sal* (*Shorea robusta*) wood was considered the best but sometimes even bamboo charcoal was used. Bellows were of different types but usually conformed to the simple bellows that one can see in operation among Indian blacksmiths even today. The large furnaces might use a pair of bellows made of bullock-hides but for the smaller ones bellows made of goatskins were enough. Some bellows were worked by hand but in some others one or two persons stood with one foot on each bellows, transferring their weight alternately from one to the other. Tuyeres were usually made of clay but sometimes bamboo poles were also used. The bellows which were worked by feet more or less conformed to the following type. First, a drum was made from a hollowed-out section of a tree trunk. Then a leather diaphragm with a hole about two inches in diameter in the centre was secured by cord around the drum. A nozzle of hollowed bamboo was fixed into the side of the drum with pitch. A wooden toggle was fitted inside the hole in the centre of the diaphragm. A cord was tied to this toggle at one end and at the other it was fixed to a bamboo pole planted firmly in the ground. A pair of bellows like this was placed with their bamboo nozzles inserted into the tuyeres. A heavy stone was put over the two nozzles to keep them in place. The bamboo poles acted as return springs to keep the diaphragms raised. The operator stood facing the furnace and kept his toes on the drums above the nozzles. He closed the hole in the diaphragm and forced it downwards by lowering his heel, whereas by raising it he caused the diaphragm to rise, drawn up by the bamboo spring. By raising and lowering the heels alternately a steady blast could easily be maintained.[22]

A common type of bellows, which can be operated by sitting on the ground, has been described by Percy.

> The bellows in common use consists of the skin of a kid or goat, taken off the animal by opening the hinder part only; the holes corresponding to the legs are sewn up; at the neck-end is inserted a nozzle of bamboo, and the tail-end is cut transversely, so as to form, when the edges are brought together, a long, straight slit for the admission of air. For a considerable length, but not to the whole extent, of each of the edges or lips composing this slit, a piece of split bamboo is firmly tied on the exterior. By this means the slit may be readily closed or opened, and made to act the part of a valve. The skin should be rendered very supple by rubbing it with oil or buttermilk. Each furnace is provided with at least two such bellows. One man is required to work them, and he sits cross-legged on the ground between them, moving them alternately in order to keep up a continuous and regular blast.[23]

[22] cf. Oldham, 1930: 13–17.

[23] Percy, 1864: 255.

In areas where the first type (i.e. of Percy's types) of furnace was used the whole enterprise was a family business with hardly any division of labour. The smelters very often used to wander around, collecting ores locally, and catered to the needs of the villages of the region. Furnaces of the second and third types were usually found in villages inhabited primarily by iron-smelters, and there the miners, charcoal-burners, smelters and blacksmiths formed distinct occupational classes.

Production statistics are difficult to obtain, but even using the first type of furnace it was possible to obtain ten to fifteen lb. of iron in a working day. We are indebted to Percy again for some of these details. In southern India, about 1844, the lumps obtained straight from the furnaces weighed about eleven lb. each and were sometimes sold for 2 annas or 3 pence each. Generally, each of these lumps yielded after forging about three lb. of iron, the best lumps yielding about six lb. The expense of forging crude lumps of iron into rough bars by hand-hammers was estimated to be 40 rupees or 4 pounds. The total expense of the bar-iron was about 80 rupees per ton, 'which a few years ago, when the above statement appeared, was less than the market price at Madras of the cheapest English bar-iron'. Four men—one master and three labourers—were required to work a furnace about 4 ft. high, and they could make only 3 lumps in 12 hours. Captain Campbell who was Assistant Surveyor-General of the Madras Establishment and who gathered these details has been credited by Percy with the following observation:

> The same writer (i.e. Campbell) asserts that the worst Indian iron he has ever seen is as good as the best English iron, and he supposes that the alleged defects of the former were due to its almost always containing a considerable portion of steel. Iron produced in his own native furnaces in Southern India could be drawn out under the hammer into a fine nail-rod not 1/16 inch thick, without splitting, and might be bent backwards and forwards 6 or 7 times before it broke. It could be twisted to a considerable extent without fracture . . . Indian iron of native manufacture is stated by some to be generally red-short, cracking at the edges when worked hot under the hammer; but according to Captain Campbell, such iron is not common.[24]

At Arnee in the Madras Presidency 100 lb. of ore (in this case washed iron-sand) yielded 33⅓ of crude iron, and 100 lb. of crude iron yielded after forging 33¼ of bar-iron. In North Canara about 200 lb. of crude

[24] Ibid: 266.

iron were obtained from each furnace at one smelting. The price was about 50 rupees per ton. In Malabar the annual outturn was estimated to be 475 tons, the selling price being about 60 rupees per ton. To produce 1lb of crude iron, 6 lb. of charcoal and 4 lb. of ore were required. The estimated profit to the smelter was about 10 rupees per ton. In Salem the cost of a furnace varied from 4 annas to 2 rupees and these furnaces were capable of containing 15 to 18 lb. of ore. This area was stated to be the principal seat of steel manufacture in south India. In Vizagapatam 100 lb. of ore yielded 25 of crude iron. In the Caroor taluk of Coimbatore, the cost of erecting a furnace (about 4 ft. high and 1 ft. in diameter) was only 1 anna 6 pice. The price of labour for 4 people to 1 furnace for a month could be 7 to 8 rupees a month. In Nagpur a furnace did not cost more than a rupee and the monthly yield per furnace was about 1 *maund* of bar-iron derived from 1¾ *maund* of crude iron. Throughout Kumaun, Garhwal, northern India and the Bengal Presidency the average price of charcoal delivered within 4 miles of the forest was about 3 annas for 30 seers (60 lb.). In Kumaun 930 seers of ore produced 327½ of crude iron or bloom metal, which in turn produced 81¾ seers of bar-iron. For 930 seers of ore, one needed 327 seers of charcoal.

J. H. Blackwell, Mineral Viewer for Bombay published in 1857 a report of the examination of the 'mineral districts' of the Narmada valley. According to him the town of Tendukhera was entirely engaged in iron-making. The details have been incorporated in Percy's work.

> The ore is a calcareous haematite occurring in limestone, about 2 miles south of the town and 4 miles from the north bank of the Nerbudda river. The workings extend over a space of 400 or 500 yards in length and about 200 in width. The ore is obtained at a depth of from 30 to 40 ft. below the surface; but it also occurs in larger or smaller quantities up to the surface. The pits are sunk every season, being all washed in by the rains; and the whole system of working is as rude and wasteful as it can possibly be. The charcoal used is obtained from hills to the north from 4 to 10 miles from the town. Every sort of wood, hard and soft, is used indiscriminately; and the jungle producing it is cut every 4 years. The charcoal is brought to the town in baskets upon the backs of buffaloes, and sold at from 3½ to 3 buffalo-loads the rupee . . . It is of very fair quality, weighing from 11 lbs. to 12 lbs. to the bushel of 2625 cubic inches. Seventy or eighty furnaces are engaged during the fine months in smelting, and iron of excellent quality is produced.
>
> The iron here . . . is divided into two sorts, *kachcha* and *pakka* . . . At Tendukhera *kachcha* iron consists of small blooms of marketable iron,

> made in a furnace more nearly resembling that of a Catalan forge than any other furnace Mr. Blackwell had seen in India. This *kachcha* iron is used for ordinary purposes . . . The *pakka* iron is made in a furnace of different description, from which it comes as a species of crude steel: it is cut in pieces, re-heated and hammered, whereby it loses its steely character and makes iron of superior quality, which is employed for purposes requiring great toughness and strength. The crude steel is used for making edge-tools, for the points of picks, crow-bars, etc. and is of very good quality. At Tendukhera the price of *kachcha* iron is from 3£. 3*s*. to 3£. 12*s*. per ton; and that of *pakka* iron from 4£. 10*s*. to 5£. 18*s*. per ton. . . .
>
> (Tendukhera) has the advantage resulting from the proper subdivision of labour, mining, charcoal-burning and smelting constituting distinct occupations. All the iron is purchased from the makers by Banians or merchants, some of whom are men of considerable wealth. The country is supplied with iron, and with articles of all sorts of iron, for a distance extending in some cases as far as 250 miles. . . . The quantity of iron made at Tendukhera . . . amounts to from 20 to 25 tons per week during 8 or 9 months of the year, all the furnaces being stopped during the rains from the difficulty of obtaining charcoal and ore.[25]

It is interesting to note that no flux was used in smelting. It is, however, reported that in the Waziri hills limestone was used as flux. A graphic account of the economic condition of the Agaria iron-smelters of the Chhotanagpur plateau comes from Valentine Ball, writing in 1880. The sponge-iron, before being hammered into a bloom, was locally called *giri*. Occasionally the Agarias themselves produced the bloom and fashioned agricultural implements and other products, but quite often they sold the giri to the blacksmiths who then made the bloom and manufactured implements:

> Four annahs, or six pence, is the price paid for an ordinary-sized *giri*, and as but two of these can be made in a very hard day's work of fifteen hours' duration, and a considerable time has also to be expended on the preparation of charcoal and ore, the profits are very small. The fact is, that although the actual price which the iron fetches in the market is high, the profits made by the Mahajans, or native merchants, and the immense disproportion between the time and labour expended and the out-turn, both combine to leave the unfortunate Aguriah in a miserable state of poverty.[26]

[25] Ibid: 268–9.

[26] Ball, 1880: 669.

III. THE METHOD OF MANUFACTURING STEEL

The classic accounts of the manufacture of wootz or steel in south India have been left by Francis Buchanan in 1807 and Benjamin Heyne in 1814. We prefer Heyne's account because of its clarity. He first saw the manufacture of steel in the Mysore territory, in a small village in the hills south-west of Chitaldurg in the Talem parganah. To convert the locally smelted iron into steel, the iron was first cut into fifty-two pieces, each of which was then put in a crucible, together with a handful of dried branches of *tangedu* (*Cassia auriculata*) and another of fresh leaves of *vonangady* (*Convolvulus laurifolia*). The mouths of the crucibles were then sealed with red mud. After this, the crucibles were placed in a circular pit with their bottoms pointing towards the centre. The pit was then filled with charcoal and lighted. The large bellows were kept blowing for six hours, which was as long as the operation took. The crucibles were then removed from the furnace, ranged in rows on moistened mud, and quenched with water. The steel was found in conical pieces at the bottom of the crucibles, whose shape they had taken. The upper or broader surfaces were often striated from the centre to the circumference.

Heyne saw steel-making in operation also at Kakerhally, a small village on the road from Bangalore to Srirangapatnam.

> Before the iron is made into steel it is heated, and reduced into pieces of eight inches length, and two inches breadth, and half an inch in thickness. It is then still so brittle that it breaks under the hammer. Is grain is coarse and white. Twenty-eight rupees' weight of it is put into a crucible, and upon it a handful of the dried branches of Cassia auriculata. This is covered with the green leaves of the Convolvulus laurifolia . . . and the opening of the crucible is closed with a handful of loam.
>
> The furnace consists of a hole in the ground about 11 ft. deep; it is one foot broad where widest and ¾ foot at the opening. The hole is filled with charcoal, and in and about the opening of it seventeen crucibles are placed; these are covered with a heap of charcoal, and bellows are kept playing on it until the contents of the crucibles are liquified . . . The operation seldom lasts longer than three hours; and is usually made four times in the course of the day, and three times in the night.[27]

The operation could vary a little from place to place. In some cases the cakes, after being taken out of the crucibles, were heated for several hours below their melting point, and turned over in the current of air from the bellows. The object of this treatment was the elimi-

[27] Heyne, 1814: 361

nation, by oxidation, of the excess of carbon in the cakes. After this the cakes were hammered into short stout bars, in which form the steel was sold. Leaves of *Asclepias gigantea* also were used in the crucible in place of those of *Convolvulus laurifolia*.

Mallet adds the following in his discussion in Watt's *A Dictionary of the Economic Products of India*: 'It appears that in some cases the blooms produced in the ordinary iron furnace, after refining in the usual way, are sufficiently steely for employment in the fabrication of edged tools, which are tempered by plunging them while hot into water. A steely iron thus made in the Jabalpur district is produced from a highly manganiferous haematite . . . But in the district of Darjeeling similar iron is smelted from magnetite containing no manganese.'[28]

An important piece of evidence in the manufacturing process of wootz was cited by David Mushet in his *Papers on Iron and Steel* in 1840. As this has not been cited by any other author as far as we could determine, we here cite Mushet in full.

> Mr. Heath has informed me that, in one particular district of India, there exists a class of workmen who practise the making of grey cast iron for the purpose of mixing in the crucible with malleable iron, as a substitute for carbonaceous matter in the formation of cast steel.
>
> This is performed in a small blast-furnace, about 8 feet high, and tapering from 18 inches diameter at the bottom to 9 inches at the top. About four times the quantity of charcoal is required than is used in the common charcoal pig iron furnace. The iron which flows out is of a grey quality, but without perfect separation, as the cinder produced contains a good deal of iron.
>
> The grey cast iron thus produced is used in a crucible with an equal portion of the malleable iron produced from the bloomery furnace for the production of wootz. This is a process which is considered to be more advantageous than the introduction of loose carbonaceous or vegetable matter along with bar iron into the crucible, as is generally practised in India. This manufacture is confined to a few families in the neighbourhood of Trinomally, and is altogether unknown to the common steelmakers at Salem, a distance of only 70 miles.[29]

A further variation in the process of steel-making was reported by Voysey from Kona Samundram in Andhra in 1832, on the basis of his 'repeated visits to the place and personal inspection of the process'. To begin with, crucibles used to be of different sizes,

[28] Mallet, 1890: 503–4.

[29] Mushet, 1840: 672–3.

according to the purpose for which the steel was to be applied, whether for the fabrication of swords, or knives, or other articles. Secondly, a piece of glass used to be put in the crucibles along with the iron matter to serve as a flux.

> The materials used in the preparation of the steel are two different kinds of iron; one from *Mirtpalli*—the other from *Kondapur*, in the proportion of three parts of the former to two of the latter. The *Mirtpalli* iron is derived originally from . . . iron sand . . . and is sent in the state of large amorphous masses of a reddish grey colour, and of an extremely porous texture. The internal fracture is often iridescent. The *Kondapur* iron is procured from an ore found amongst the iron clay, at a place about 20 miles distant. It is said to be of a dirty brown colour and very fragile. The iron, however, is moderately compact and of a brilliant white fracture. Occasionally it contains some ingredient which spoils the steel, rendering it excessively brittle: the natives assert that the adulteration is copper, but it is more probably arsenic. The mixture being put into the crucible, the fire is excited and kept up for 24 hours. It is then allowed to subside, and the crucible is taken out and placed on the ground to cool. When quite cold it is opened, and a cake of steel of great hardness is found, weighing on an average about a pound and a half. The cake is covered with clay, and annealed in the furnace for 12 or 16 hours. It is then taken out and cooled, and again annealed, and this may be repeated a third or fourth time until the metal is rendered sufficiently soft to be worked.[30]

In 1899, while commenting on S. A. Bilgami's paper, 'On the Iron Industry in the Territory of His Highness the Nizam of Hyderabad, Deccan', Cecil Von Schwarz offered a description of the steel-making process which must be considered unique in the sense that in this description a mixture of iron ores ('a mixture of magnetic sand and laterite') was put inside crucibles in place of a quantity of already smelted wrought iron. In this case steel was being obtained directly from the ores. Schwarz was the superintendent of the government iron works in India and thus he surely knew what he was writing about. Considering the unique character of this report we here cite Schwarz in full.

> The following is a brief repetition of a description of this process which I discovered n some old records. . . . The smelting process was carried on in flower-shaped crucibles made of fire-clay kneaded well with cow-hair and oil, then dried and baked. Each crucible was provided with a bell-shaped cover having a small opening on the top to be closed with a stopper made of fire-clay. A bit of glass slag was put on the bottom of the crucible,

[30] Voysey, 1832: 246–7.

> *after which it was filled up to two-thirds with a mixture of magnetic sand and laterite* (italics ours), then covered with leaves... and mixed with some charcoal powder, the latter acting as a reducing and carburising ingredient. The crucible was then placed on an underlayer consisting of a disc-like bit of fire-clay, the cover with the stopper put on, and the whole buried in a heap of charcoal. A gradually increasing heat was now kept up for from about 18 to 20 hours with four bellows when the melting commenced. The master controlled this process most carefully by lifting the stopper of the crucible cover and examining from time to time the contents, at the same time stirring the liquid mass to make it homogenous. As soon as the proper degree of liquid condition was reached, which moment was carefully controlled, the bellows were stopped and the crucible with its contents allowed to cool down slowly. The crucible was then broken and a cake of steel of about 3½ to 4 lbs. in weight obtained. The cake of steel was, however, much too hard, and in order to reduce its hardness to the proper degree it was subjected to a kind of tempering process. After the cake was carefully cleaned from adhering slag, etc. it was dipped into a mixture of iron ore and manganese ore powder with clay water, then dried and exposed to a light red heat. This tempering process was repeated until the master, by means of chisel and hammer, was convinced that the steel had the exact correct degree of hardness. It was then called 'kurs', and was sold to Persian merchants at the rate of one to one and a half rupees per seer (about 2 lbs.) who transported this material on mules' backs to Damascus, thus furnishing the principal material for the celebrated Damascus blades of the Middle Ages. The excellent quality of this material has never since been reached..[31]

Schwarz calls this steel 'crucible cast steel' and he is emphatic that 'the natives used iron ore as a raw material'. He also explains that the perfect hardness of this steel was obtained 'by introducing into their cast steel an excess of carbon and by taking this excess gradually away afterwards by means of the slow tempering process, having it thus completely in their power to attain the exact degree, by interrupting this decarburising process exactly at the proper time in order to get a cast steel of a quality exactly suitable for the purpose'.[32]

IV. SOME REGIONAL DETAILS

Ferozepur

An anonymous report (initialled A. E.) was published on the iron-

[31] Schwarz, 1899: 97–8.
[32] Ibid.

smelting industry of Ferozepur in the Gurgaon district of modern Haryana in 1831. Two kinds of ore were used: the Narnoul ore of two varieties (one is nodular and the other is found in 'large masses') and the Ferozepur ore found in *Kankar*-like concretions. The Ferozepur ore used to be called *Buri*. These ores gave a return of only 15–25 per cent. The Narnoul and Ferozepur ores were smelted together but they were first broken into small pieces ('about the size of a child's marbles') and then mixed with an equal quantity of charcoal which was bought in the neighbourhood for 25 rupees per hundred maunds. A wall with two doorways was built at the back of the furnace and supported a shed or lean-to under which the bellows-men sat and plied their task day and night. Two sets of men were employed for the purpose. The furnace (locally called *Mandari*) was merely a square receptacle with two openings—one at the top for the charge of the ore, and the other in the lower portion for drawing out the finished lump and molten slag. A hollow was sunk in the ground to receive the dross or molten slag. Two bellows were used for each furnace and the nozzles were inserted very low down at the back, opposite the lower external opening. The bellows were made of a single skin from an unusually large goat.

After the operation was complete, the lump of half-melted iron was taken out by means of large iron pincers (called *Sangasi*), with proddings by crow-bars through the bellows-hole from behind. Heavy wedge-headed hammers were used to trim away all superfluous knots and lumps of impure metal from this mass. It was then heated in a forge (called *aran*) and broken into smaller pieces. These pieces were forged into bars 'about a cubic in length and two fingers thick in the middle, tapering at both ends'. This iron used to fetch 4 rupees per maund.[33]

Kathiawar

On 5 December 1840, Captain Legrand Jacob read a 'Report on the Iron of Kattywar, its Comparative Value with British Metal, the Mines and Mode of Smelting the Ore' in the Royal Asiatic Society, London. According to him, iron ore was chiefly found in the north-western part of the peninsula. Six 'foundries' (or two or three more) used to run at any one time 'during the fair season' in this region, and he visited two of them, those at Ranawao and Ranpoor in the Rana's and Jam's taluks respectively.

[33] Anonymous, 1831.

The location of iron-smelting furnaces at these places was determined by their proximity to the Barda hills where both charcoal and ore were easily available. The ore was mined in the hills by digging circular pits to a depth of 5–20 ft. and sifted, washed and despatched to the foundries in carts or on bullocks and donkeys. The smelting furnace was long and narrow and covered by masonry lined with clay to keep in the heat. There were two openings on two sides of the furnace; one was meant for the working of the bellows and the other was intended to let out the scoria. An alternate charge of charcoal and ore was applied and the bellows were two pairs to each furnace. They were large bellows made of bullock hides sewn round bamboo hoops in vertical rings. The outturn was about forty per cent of the ore. There were two smelting operations in a day. The lump from the furnace was heated in a forge and broken into pieces which were then wrought into small bars for the market. Two varieties of iron were produced—*chontia*, which was the inferior variety (produced inferior ore?) and *marka*, which was the better sort. Jacob calculated that the daily profit from one foundry was about 15 cowries or 5 Ahmedabad rupees. The daily pay of a workman was one cowric.

> The workmen commence their daily toil at the first dawn of light, and cease generally a little before sunset; they appeared to labour with much perseverance and industry; it is difficult to witness, without pain, the struggles of these poor people for a subsistence, which our superior skill is yearly rendering more arduous. The annual produce of one foundry is about sixty-five Bombay khandies, or between sixteen and seventeen tons; and taking the number in the Peninsula at six constantly working through the fair season, and two or three more occasionally, the amount of iron fabricated yearly in Kattywar cannot be fixed at much above a hundred tons annually, and, may be safely estimated under 150 tons at the outside.[34]

South India: Francis Buchanan's Testimony

As far as one has been able to determine, Buchanan in his report in 1807 described two main types of iron-smelting furnaces in south India, especially in Mysore and Malabar. In both cases the description is very specific, and is aided by sketches.

Type 1: 'The smelting furnace is made in the front of a square mound of clay . . . In the front, the mound is twenty-two inches high, and three feet broad. In this, from top to bottom, is made a semi-cylindrical cavity, about a foot in diameter. On the ground, in front of the

[34] Jacob, 1843

cavity, is laid a stone six inches high, a foot long, and a foot broad. Contiguous to this is placed another stone a foot square, and two inches thick. On the top of this is fixed a small piece of timber, behind which rises another mound of clay, sloping upwards gradually, and widening as it recedes from the furnace. On this rest the bellows, of which there are two. Each consists of a whole buffalo's hide, removed without cutting it lengthwise. Where it has been cut at the neck, it is sewn up, so as to leave a small opening for a wooden muzzle, which is made fast to the piece of timber before-mentioned. The hinder part of the skin is slit vertically, and the one side is made to lie over the other. In the middle of this outer side is fastened a ring of leather, through which the workman passes his arm, and seizes the upper angle of the skin, which serves as a handle. When he draws back his arm, the opening in the hinder part of the skin is dilated, and admits the air; when he forces his arm forward, the opening is closed up, and the air is forced through the muzzle. The lower part of the bellows is retained in its place by a rope fastened to the lower angle, and supported by an elastic piece of timber, which is fastened to one of the posts of the hut, like a turner's lathe. The muzzles of both the bellows are inserted in one common tube, which is made of baked clay, and is placed in a sloping direction, so as to pass through a mass of moist clay, that occupies the front of the furnace above the first mentioned stone. Above this is placed a large tile; and the empty spaces between this and the mound (1) are filled up with moist clay. The furnace is now cylindrical, and open at the top, on which is placed a chimney made of baked clay. . . .'[35]

Type 2: 'The furnaces are excavated out of the front of a mound of clay, which is 4 feet high behind, and 5 feet four inches before; and about 7 feet wide, from front to back. The excavation made for each furnace is 2 feet 11 inches wide, and 2 feet deep; and is dug down from the top of the mound to the ground. From behind, opposite to each furnace, an arched cavity is dug into the mound; so as to leave a thin partition between the two excavations. For allowing the vitrified matter to run off, there is in this partition a hole one foot in diameter. Above the furnace is erected a chimney of clay, built with four plain sides, which in two different places is strengthened by four *Bamboos*, lashed together at the angles. The front of the chimney consists of baked clay, two inches in thickness. Behind, the clay is gradually thickened toward the summit; so that the upper mouth of the chimney

[35] Buchanan, 1807, 1: 171–2.

is contracted to 8 inches in depth by 2 feet 11 inches in width. The front of the furnace is quite open.

Early in the morning, when going to smelt, the workmen put wet sand mixed with powdered charcoal into the bottom of the furnace; so as to fill it up as far as the hole in its back part, through which the vitrified matter is to run out. The sand and charcoal are well beaten, and formed so as to slope from the outer and upper edge, both toward the hole and toward the ground in front of the furnace. The hole is then well stopped with clay; and clay pipes are inserted at each corner of the furnace, for the reception of the muzzles of the bellows. A row of clay pipes, eight or ten in number, is then laid on the surface of the sand, at right angles to the back of the furnace. Their outer ends projects a little beyond the front, and their inner ends reach about half way to the back. The front of the furnace is then shut up with moist clay; and stoppers of the same are put in the outer mouths of the pipes. By removing these stoppers, and looking through the pipes, the workmen judge how the operation is going forward.'[36]

These two types of furnaces seem to be distinctive of south India. The operational details, however, did not differ significantly from those in other parts of India, although in these parts one notices the dominance of black sands as iron ore. The collection of this type of sand and its preparation as ore used to be undertaken in the monsoon months.

> During the four months of heavy rains, four men are able to collect as much sand as a furnace can smelt in the remainder of the year. In order to separate the earth and sand, which are always mixed with it in the channel of the torrent, it requires to be washed. These men get ten *Fanams*, or 6 *s*. 8½ *d*. a month, and the nature of their service is similar to that of the farmers' servants, being bound by occasional advances of money to continue in the employment of the master. During the remaining eight months of the year they work at the forge.[37]

The lump, after being taken out of the furnace, was broken into pieces and then forged into bars. It appears that people of particular caste groups were involved in the smelting activities. At one place Buchanan writes that 'the operation is performed by *Malawanlu*, the Telinga name for the caste called Parriar by the natives of Madras'[38]

[36] Ibid: 437–8.
[37] Ibid: 171.
[38] Ibid: 29.

while at another he states that 'it is wrought by the low people called *Siclars*'.[39]

There were also variations in the pattern of collection of ore and charcoal, as the following extract will show. 'The landlords in general prepare the ore by their own slaves, and sell it to the smelters ready for the furnace. The people who make the charcoal pay a trifle to the landlord for permission to carry on their business.'[40]

Buchanan has also tried to list the expenses and profits of the iron forges. These also throw light on the organizational details of iron-smelting in the south in the beginning of the nineteenth century.

> The expense that attends the working of one of these iron forges is as follows:

	Fanams
To 4 men for collecting iron sand, at 10 *Fanams* each for 4 months - - - -	160
To 6 men to make charcoal, 4 for the smelting-house, and 2 for the forge, during 8 months, at 8 *Fanams* monthly for each - - - -	384
To 4 labourers at the smelting-house, for 8 months, at 10 *Fanams* each - - - -	320
To 6 labourers in the forging-house, of whom 1 has 12 *Fanams*, the others 6 *Fanams* a month, for 8 months -	336
To the government paid yearly; for making charcoal 60 *Fanams*, for ground rent for furnace 20 *Fanams*, for ditto for servants' houses 20 *Fanams* - - -	100
Fanams	1300

> The smelting-house burns thrice a day, for about eight months of 32 days each, without any allowance for holidays, and at each time produces as much iron as, when forged, sells for from two to three *Fanams*.[41]

At two *Fanams*, the profit was calculated to be 236 *Fanams* per year whereas, at three *Fanams*, it was estimated to be 1004 *Fanams* per year.

Buchanan, however, does not specifically dwell on the quantity and price of steel produced in this region. At one place, after a description of the iron-smelting process, he simply states: 'the same persons also make steel'.[41] He then goes on to describe the

[39] Ibid, II: 283–4.
[40] Ibid: 495.
[41] Ibid, I: 175–6.

steel-making process but does not dwell on its commercial aspects. According to Heyne, however, one hundred pieces of steel, products of roughly a hundred crucibles, 'are sold for four Cantaray pagodas, i.e. fifteen pounds; cost about seven shillings'.[42]

The Khasi Hills

A smelting furnace described by J. D. Hooker in his *Himalayan Journals*, volume II (1854), and cited by Percy on that basis does not conform to the usual type of smelting furnace.

> There is neither furnace nor flux used in the reduction. The fire is kindled on one side of an upright stone (like the headstone of a grave), with a small arched hole close to the ground: near this hole the bellows are suspended; and a bamboo tube from each of its compartments meets in a larger one, by which the draught is directed under the hole in the stone to the fire. The ore is run into lumps as large as two fists, with a rugged surface: these lumps are afterwards cleft nearly in two, to show their purity.[43]

It is important to note that the 'hole' here is not in the ground; the upright stone must have a hole through which the blast has to be directed at the fire on the other side of the stone, in its shadow, so to speak. Secondly, the bellows are set up by tying them to a cord and hanging it from a tree branch or a house beam. A person stands on the bellows and operates it 'by a wriggling motion of the loins, and the strength of the leg'.

The Orissan Uplands

A note by H. F. Blanford, which has been cited by Percy, describes the iron-smelting activities in this region. First, the iron-smelters belonged to some aboriginal tribes which in Talcher and the neighbouring districts were known as the Kols. 'They are to a certain extent nomadic in their habits, remaining in one spot only so long as plentiful supplies of ore and wood are obtainable in the immediate vicinity. These failing, or, as frequently happens, on the occurrence of anything that is regarded as of evil omen by these superstitious communities, they transport themselves and property to some more propitious site, and work away as before.'[44] Their furnace made of local earth and generally strengthened by a sort of skeleton of strips of flexible wood, was about 3 ft. high, its mean internal diameter being about 1 ft. The sponge iron was extracted after smelting

42 Heyne, 1814: 361.
43 Percy, 1864: 261–2.
44 Ibid: 264.

through the front opening, and during the smelting it was plastered up with the nozzle of the bellows inserted at the bottom. The bed of the furnace was slightly inclined towards the slag-hole on one side of the furnace to allow the slag to flow freely into a small trench made for this purpose outside.

Iron-making at Mirjati, Chhotanagpur

Mirjati is the name of a village in the foothills of the Dalma range, near the modern iron and steel works in Jamshedpur, Bihar. In January 1918, Andrew McWilliam, a metallurgist of Sheffield, went to this village from Jamshedpur to study the iron-smelting operation undertaken by a family of smelter-blacksmiths—Bidhan Kamar and his two sons and a daughter. The total operation lasted from 8 a.m. to 1.35 p.m. The ore used was a brown haematite, found in comparatively small masses near the village. Charcoal was made by the family in the adjacent jungle free of charge. No flux was applied. The ore and the charcoal were first reduced to the size of beans. Two baskets (14 ins. in diameter and 12 ins. deep) of bean-sized ore required four baskets of charcoal of about 8 seers (16 lb.) each and produced about 5 seers (10 lb.) of iron. The furnace, made of local clay, was 3 ft. 10 ins. high from the ground level. At the ground level the external diameter was 2 ft. 6 ins.; at 3 ft. 4 ins. above it was 1 ft. 7 ins., but at the top it was slightly greater—1 ft. 9 ins. The top was flat and level to hold a charge of ore and charcoal. The internal diameter sloped from 4½ ins. at the top to 15 ins. at the bottom. There was an arched opening in the front from ground level upwards, 15 ins. broad by 14 ins. high. This was plastered up during the operation with a suitable hole at the bottom for the insertion of the nozzle of the bellows. There was another small hole at the side which was generally closed with clay and opened occasionally to tap off the cinder. The bottom of the furnace inside was in the form of a rounded cavity below the ground level. That was where the particles of iron might drop and coalesce, below the level at which the blast entered. The bellows were hollowed-out segments of tree-trunks with buffalo-hides tied securely to them. The bellows were operated by feet. When enough iron was made, the spongy mass was taken out by breaking open the temporary front and was immediately cut in two with an axe. One-half was reheated in a charcoal fire and forged into a piece about 1½ ins. square. The two pieces together weighed 7¼ seers (14 lb. 8 oz.). About the family, McWilliam wrote: 'They make ploughshares, sickles, arrow-heads, axes, knives, and all domestic articles they require. On the average the family run one heat a week, taking the remainder of the time to attend to their farm, gather ore and wood, make charcoal, and make iron into articles for sale.'[45] They could sell all the articles they made without leaving home.

[45] McWilliam, 1920: 162–3

CHAPTER SIX

Summary and Conclusions

REVIEW OF RESEARCH

We began this volume with a general history of research on Indian iron. This history of research is of absorbing interest not merely because it dates from as early as the eighteenth century but also because many streams of thought and approach merged into it. European interest in Indian iron dates from the Graeco-Roman period, or more correctly, the Hellenistic-Roman period. Imperial Rome had an eastern iron trade, and much earlier, Alexander had received a gift of Indian iron. This interest was possibly greatly strengthened in the Middle Ages when the steel used in the manufacture of the famous sword-blades of Damascus used to come from India. The fame of Indian iron and steel in Europe persisted possibly till the threshold of the Industrial Revolution in the second half of the eighteenth century. This is clear from two references which we have cited and which belong to an earlier period. In his presidential address on Indian iron to the Staffordshire Iron and Steel Institute in 1911, I. E. Lester mentioned the tradition whereby the Westphalians used to 'roll puddled steel into ½ inch or ¾ inch squares and sell it in Hamburg under the name of Indian steel'. This tradition can belong only to the period before the Industrial Revolution and shows that Indian iron used to enjoy considerable fame in Germany of that period. The second reference relates to the purchase of ingots of Indian steel by the French Consul in the bazaars of Cairo in 1722. This steel was known to be the raw material of the Damascus blades and 'rated most highly in Egypt'. We know from H. Voysey's report that the traders from Iran used to come to certain areas of Andhra to purchase pre-industrial Indian steel as late as the early years of the eighteenth century. It is not, however, known that this trade included Egypt and possibly Europe too. Much too little is understood about the trade in Indian steel in the pre-modern period.

The scientific curiosity in wootz was short-lived. George Pearson's experiment in the Royal Society was in 1795 and, with David Mushet's analysis in 1805, the process of wootz-manufacture came to be well understood. James Stodart in 1814 possibly wanted to find the ways in which wootz could be best used. The experimental interest was pursued by Michael Farraday (1819) and James Breant (1823). One of the results of this scientific curiosity was that the superiority of wootz to anything Europe had produced till that period came to be well established. The attempts to understand the manufacturing details of the Damascus sword-blades, which one notices in the writings of Henry Wilkinson (1836), Colonel Anossoff (1843) and James Abbott (1848), were related to this scientific curiosity. These details obviously generated a great amount of manufacturing and commercial interest in the first half of the nineteenth century.

India's importance as a major and early centre of iron production clearly emerged with the geological enumeration of her iron ores, and the miscellaneous government reports on the pre-industrial iron-smelting operations and the production of iron in different parts of the subcontinent. The major types of pre-industrial iron-smelting operations were systematically studied first by John Percy (1864), but the most comprehensive survey of its kind, both in the field of ores and in the methods and distribution of pre-industrial iron-smelting, was undertaken by Valentine Ball (1881) in the volume on economic geology as part III of *A Manual of the Geology of India*. His name is not very well-known in Indian archaeology but Indian archaeologists have a lot to thank him for. A geologist of the Geological Survey of India, Ball primarily worked in eastern India, especially in the Chhotanagpur plateau region; and he left behind, apart from his geological researches, a large number of archaeological writings which are still useful to the archaeologists working in that region. His survey of the different metallic ores and economically useful stones was on the subcontinental scale, with a detailed listing of the sources. I take this opportunity to record my deep sense of indebtedness to Valentine Ball.

In any case, the pre-industrial iron-smelting operations in different parts of India made a number of metallurgists interested in this problem, and one comes across a number of writings by metallurgists on this topic in the later part of the nineteenth and the carly part of the twentieth centuries. As far as the history of iron is concerned, the foundation was laid by a Chemistry teacher, Panchanan Neogi, who systematized in 1914 the geological, ethnographical, archaeological, metallographic and literary data available till that date. In the field of metallographic analyses, the most significant attempt till that period was by Robert Hadfield in the context of a few dated iron objects from Ceylon. Neogi offered a pioneering synthesis of the entire range of data. If anything, the present work is in the academic tradition of Neogi. There was no specific concern with the beginning and history of iron in India in Indian archaeology till 1950, and although the credit of focusing attention for the first time on this issue goes to Colonel D. H. Gordon, it must be pointed out that his approach to the problem was entirely diffusionist. In this he set a tradition which found its most detailed expression in N. R. Banerjee's *The Iron Age of India* in 1965. Banerjee's discourse on such ponderous issues as the Aryan homeland and the Hittite monopoly of iron-smelting is an excellent example of labyrinthine scholarship. His arguments were, however, not seriously questioned till 1976.

Eventually, a proper history of the use of iron in India will have to be primarily a technological history, based on the properly planned metallographic studies of ancient iron objects from different areas and different periods. A planned regional study of this kind has been undertaken in recent years by Thelma Larson Lowe who pursued a project entitled 'Archaeological Field Survey in Northern Andhra Pradesh—The Identification and Distribution of Iron-smelting and Crucible Steel Production Sites' between 1987 and 1989 in the Nizamabad district, western Karimnagar district and southern Karimabad distict of Andhra Pradesh.[1] Basically ninety-three deposits of iron industrial debris on seventy-four sites have been identified, and on thirteen of them, wootz had been produced by what is called the Deccani process. The research results are still unpublished, but one hopes that this will throw a lot of light on the iron-manufacturing process, especially the wootz-manufacturing process, in a part of India. However, till more works of this kind are attempted, the significance of a comprehensive synthesis of the geological, archaeological, metallographic, literary and ethnographic data on Indian iron cannot be ignored. We have tried to offer in this volume such a synthesis.

[1] Lowe, 1989.

DISTRIBUTION OF ORES

Our review of the distribution of iron ores suitable for pre-industrial smelting has underlined the ubiquitous character of its spread outside the major alluvial stretches. To begin with, this distribution covers the entire mountain system to the north, right from the Kirthar zone in Sind in the west to the hills of upper Assam in the east. The Baluchi hills, the Northwestern Frontier, the Panjab and U. P. Himalayas, the northern hills in West Bengal, and the Khasi-Jaintia hills of modern Meghalaya all fall in between these two points, making this vast arcuate region a major belt of pre-industrial iron-smelting in the subcontinent. In the peninsular landmass to the south of the Indo-Gangetic-Brahmaputra alluvium, iron suitable for pre-industrial smelting occurs virtually everywhere. Not many of these deposits are of modern commercial value but as we have already noted (chapter 2), 'the primitive iron-smelter finds no difficulty in obtaining sufficient supplies of ore from deposits that no European iron-maker would regard as worth his serious consideration'. This is a very important point to remember which makes nonsense of many learned speculations in this field. For instance, D. D. Kosambi argued that the might of the imperial Mauryas was based on the utilization of the rich iron ore reserves of south Bihar,[2] whereas the truth is that the Mauryas could get any amount of iron ore suitable for pre-industrial smelting from many parts of their large empire. Incidentally, the rich iron ore reserves of south Bihar are situated in the fairly dense and inaccessible (till the modern period) forested region of south Bihar and there is no archaeological, or even circumstantial, evidence of Mauryan penetration into this area. The same is true of H. C. Bharadwaj's notion that the ore used for early historic Rajghat iron objects came from Bihar.[3] The deposits in the neighbouring Mirzapur district are surely a more plausible source. In fact, we feel that the iron ore deposits in the neighbouring hilly terrain played a major role in the development of the iron industry in the Gangetic valley. R. C. Gaur apparently understands this point when he writes: 'The main source of iron for the P.G.W. people of Atranjikhera was most probably the region extending from south of Agra to Gwalior containing rocks in which iron content is quite high. It would be worthwhile to mention that extensive deposits of banded haematite-

[2] Kosambi, 1965.
[3] Bharadwaj, 1973.

quartzites found in these rocks are considered important and have been worked out on large scale by indigenous smelters.'[4]

An important point to be noted in this context is the availability of a number of ore-types in the same area. This leads us to speculate whether there was any preference for a particular ore-type in the areas with different varieties of ore. The answer is not easy to obtain but it is logical to assume that different ore-types could be used by the smelters of an area and that they were likely to show preference for the easily accessible varieties. The Agarias of Palamau are known to have used both magnetic and haematite ores for their iron-smelting operations, but Valentine Ball points out the following:

> The iron-ores to which my attention was particularly directed in Palamow admit of a triple classification, founded both on their geological relations and age, and on their chemical composition. In the first class there are magnetites or magnetic ores which occur in the crystalline and metamorphic rocks. In the second class there are siderites and haematites which are found in the coal-measures. And in the third class there are haematites which are found in the laterite. The magnetite is of great purity and excellent quality; but its abundance in any one spot in suitable form for extraction is very doubtful. The haematites of the laterite, being situated on the tops of lofty plateaus upwards of 300 feet high, are practically inaccessible, though the quality of the richer varieties leaves nothing to be desired. It is to the ores of coal-measures, therefore, the siderites or iron carbonates, and haematites or iron oxides that I have been compelled to shew preference, as being most likely to afford an inexhaustible and easily worked supply of fairly good ore.[5]

Incidentally, the wide distribution of iron ores in the subcontinent discounts the possibility of long-distance trade in this material. For the alluvial areas far from the known deposits of iron there could have been trade in iron 'blooms' or, possibly, even in iron products. There is no reason why the situation at Tendukhera in the Narmada valley, as reported in the middle of the nineteenth century, could not have been common in the context of ancient India as well. 'All the iron is purchased from the makers by Banians or merchants, some of whom are men of considerable wealth. The country is supplied with iron, and with articles of all sorts made of iron, for a distance extending in some cases as far as 250 miles.'[6]

[4] Gaur, 1983.
[5] Ball, 1880.
[6] Percy, 1864.

ARCHAEOLOGICAL DATA

In our chapter on the nature of the archaeological evidence we put forward the basic data on the excavated iron objects in different parts of the subcontinent. In Baluchistan we discounted the possibility of an early date for the cairn-burials of the region. We argued that these cairn-burials could not have been earlier than the first century B.C. and would have continued possibly up to the seventh century A.D. The only genuinely early occurrence of iron in Baluchistan was at Pirak (800 + B.C.). Although only arrowheads have been reported, there is a blacksmith's furnace at the site, showing that the objects were made locally. The exact nature of the beginning of iron in the Gandhara Grave Culture is not clear but this falls around the opening of the first millennium B.C. The recorded tool-types may be considered as belonging to the first half of the first millennium B.C. As Jettmar points out, this is a span of time also suggested by the cheek-bar of a horse from Timargarha. The suggested central Asian-Afghanistan links of the Gandhara Grave Culture are beyond dispute, but barring the example of the Timargarha cheek-bar, the iron objects of this culture are typologically quite common—spearheads, arrowheads, pins or nails, etc. Little is yet known about the occurrence of iron in the megalithic level of Gufkral in Kashmir, but a date around 1000 B.C. has been suggested for this level. A date in the first half of the first millennium B.C. is also possible for the early iron at Saraikhola. The evidence from the region as a whole, including Kashmir, seems to be fairly consistent.

The evidence from the period IV at Balambat, Bhir mound and the early levels of Charsada falls in the second half of the first millennium B.C. and includes a wide range of objects including elephant goads and scale-pans. The diversity of types during this period prepares us mentally for the very wide range of iron objects from the Saka-Parthian Sirkap. These Sirkap objects are dated in the period from the first century B.C. to the first century A.D. About 40 basic types have been listed, and one notes the occurrence of a large number of identical tool-types from roughly the same period at Shaikhan Dheri. The evidence from the later historical contexts is generally unknown in many areas and that is why we have cited the details of an excavated smithy in the period V (sixth-tenth centuries A.D.) at Damkot in the same region. The data from the north-western region as a whole are thus both comprehensive and detailed and here we may underline only the following areas of uncertainty: the cultural links of the iron-bearing levels of Pirak, the nature of the beginning of iron in the Gandhara Grave Culture and finally, the details of the iron-using megalithic culture in Kashmir.

It is interesting that there is virtually no early historic evidence of iron in the Panjab plains and Sind. This, of course, only underlines the lack of excavations of early historic levels in these areas.

The extensive occurrence of iron in the Painted Grey Ware culture of the Indo-Gangetic divide and the upper Gangetic valley is now proved by the evidence from Atranjikhera, Hulas, Jakhera, Hastinapur, etc. The Sheer diversity of iron objects in the Painted Grey Ware level at Atranjikhera (16 categories of finds, ranging from blacksmith's tongs, clamps and eyed needles to arrowheads, spearheads and chisels) suggests the development of a full-blown iron industry in the upper Gangetic Valley during this period. Hoes and sickles occur in the 'mature PGW culture' at Jakhera. We find the division of the Painted Grey Ware period at Jakhera into two sub-periods very interesting. In Period IIA this pottery is said to occur only in the upper levels but the period as a whole is said to contain iron slag and bloom. We have argued that this suggestion in the case of Jakhera and the reported occurrence of iron in the black-and-red ware deposit at Noh in the Bharatpur area of Rajasthan perhaps imply that the beginning of iron in upper Gangetic valley was earlier than the Painted Grey Ware deposit. In any case, the iron industry of the Painted Grey Ware period in the upper Gangetic valley is an extensive and well-rooted industry, the roots of which have to be earlier than this period, either outside the valley or in the valley itself. We have also argued that the date of 1100 B.C. which was suggested by B. B. Lal as the initial date of the Painted Grey Ware at Hastinapur is still an acceptable date for this ware in the upper Gangetic valley.

In the succeeding Northern Black Polished Ware level the site which has shown the widest range of evidence is again Atranjikhera where there are 31 categories of finds in this level. The evidence is more or less supported by the iron finds from Rupar, Hastinapur, Kausambi, etc. The evidence from this region as a whole is both comprehensive and detailed, but at the same time, the issue whether iron began in the black-and-red ware deposit in the upper Gangetic valley needs further examination in the field.

In the middle Gangetic valley and the adjacent Chhotanagpur region, the earliest association of iron is with the black-and-red ware which is 'chalcolithic' in origin but occurs also in association with iron. At two sites—Mahisdal on the bank of the Kopai in West Bengal and Barudih on the bank of the Sanjay in Singhbhum in Bihar—iron is found associated with microliths and neoliths respectively. The dates in both cases are 1000 + B.C. The tool-types from Pandu Rajar Dhibi, Mahisdal, Chirand, Barudih, etc. suggest a well-developed and extensive iron industry, and what is remarkable is that in the column of the black-and-red ware in this region iron simply begins to occur in its upper phase without bringing about any change in the material equipment of the earlier chalcolithic level. In the context of this region we would point out that the non-publication of the total excavated evidence from Chirand, the most significant site in the entire region, is a major handicap in our understanding of the archaeological sequence of the middle Gangetic valley. In the context of eastern U.P. we need more information on iron in the black-slipped ware deposit at such sites as Ganwaria in the Basti district. In this connection we may also note that the Chhotanagpur plateau portion of this region must have been the major source of iron for the valley. We may speculate that in the Chhotanagpur plateau there were important production centres (cf. Saradkel) for the supply of iron objects and/or bloom to the valley.

Further to the east, there is no early archaeological evidence of iron in the Brahmaputra valley or Assam. The same is more or less true of the adjacent Bangladesh where there are two reported early historic sites—Mahasthangarh on the bank of the Karatoya in the Bagura district, and Wari-Bateshwar on the bank of an old course of the Brahmaputra in the Narsingdi area of the Meghna delta. One is not quite sure of the Mahasthangarh archaeological sequence but the Northern Black Polished Ware and Sunga terracottas are known to occur at the site. Wari-Bateshwar has not been adequately explored but is known to possess a large number of silver punch-marked coins, among others. From the present point of view what is important at this site is that it has yielded on the surface about five hundred heavy iron objects (5–6 ins. long, 3 ins. thick, and 2–3 kg. in weight) which can be interpreted only as imported iron bloom. The Bangladesh National Museum has a few of these specimens on display as 'iron handaxes' (!)

The Assam situation is no doubt perplexing but no early historic archaeological level has yet been isolated in Assam. The hills fringing the valley are, as we have already noted, rich in iron ores and possess a pre-industrial smelting tradition. In Rajasthan the early historic iron objects showed a wide range of types at Bairat, Sambhar, Rairh and Nagri, but in this region perhaps the most important problem is to determine if the black-and-red ware deposits at Noh and Ahar really contained iron. This has been frequently doubted (although mostly in conversations among archaeologists) but we feel that there is no substantial reason to doubt this occurrence. As we have argued, the Ahar iron objects are spread in five trenches and belong to different layers, and cannot be interpreted in any way as belonging to disturbed layers. The chronology of Ahar iron is also very interesting, going back (after consensus calibration) to the first quarter of the second millennium B.C.

The early historic data from Gujarat are still limited but Harappan Lothal yields one object of iron. The evidence of iron in the pre-Northern Black Polished Ware black-and-red ware deposit at such sites as Nagda, Eran and Dangwada in Malwa is quite explicit. The debate is mainly about the chronology of this level. There is no appreciable time-gap between the end of the 'chalcolithic' black-and-red ware deposit and the iron-bearing black-and-red ware deposit at Nagda and elsewhere and, on this basis, we date the beginning of iron in Malwa soon after the middle of the second millennium B.C. Nagda yielded the largest number and varieties of early historic iron objects—210 specimens grouped into 14 categories. The iron objects from the megalithic context of Vidarbha in Maharashtra seem to begin in the first quarter of the first millennium B.C. and are, as a group, affiliated to the iron objects from the south Indian megaliths. It is a rich iron industry with about twenty types of objects.

The occurrence of iron at Prakash and Bahal in the Khandesh area of Maharashtra should be contemporary with the iron in the same context at Nagda, Eran, etc. In the early historic level of Maharashtra, Nevasa seems to be the richest site as far as the number of excavated iron objects is concerned. The south Indian megalithic iron objects have been discussed with reference to their twenty major types, following the classification of Allchin, and also with reference to the excavated iron objects from the site of Adichanallur. The beginning of the south Indian megalithic iron seems to fall in the closing centuries of the second millennium B.C. The only intelligible data from Orissa have been found in the early historic level of Sisupalgarh.

The main regions which have been discussed in this section with reference to their major iron-bearing sites and cultures are shown in the following list.

1. *Baluchistan*: Pirak (800 + B.C.), cairn-burials (C.first century B.C.-seventh century A.D.).
2. *The Northwest*: the Gandhara Grave Culture (first half of the first millennium B.C.), Saraikhola (first half of the first millennium B.C.), Balambat IV, Bhir mound IV, Charsada (Achaemenid?), Bhir Mound II (Mauryan), Sirkap III–II, Shaikhan Dheri (Saka-Parthian-Kushan) and Damkot (later history).
3. *The Panjab plains and Sind*: no positive data but Tulamba III (C. eighth–twelfth centuries A.D.).
4. *The Indo-Gangetic divide and the upper Gangetic valley*: Atranjikhera, Hastinapur, Alamgirpur, Hulas and Jakhera (Painted Grey Ware in the *Doab*, C.1100 B.C.-) Jodhpura (Painted Grey Ware in Rajasthan), Kausambi (the easternmost section of the *Doab*, Painted Grey Ware) and Ahichchhatra (Painted Grey Ware, also in the *Doab* but the evidence is very limited at this site); the early historic level of this region is represented in our discussion by Rupar, Hastinapur, Kausambi and Atranjikhera.
5. *The Middle Gangetic valley and the adjacent areas*: Ganwaria (black-slipped ware deposit), Kodihawa (black and red ware), Panchoh (handmade and ill-fired corded and plain red wares), Sonpur, Chirand, Taradih, Pandu Rajar Dhibi, Mahisdal, Bharatpur, Bahiri, Barudih, etc. (black-and-red ware deposits) in Bihar and West Bengal. The early historic evidence in Bihar is discussed with reference to Pataliputra, Saradkel and Karkhup. A reference has also been made to the numerous iron blooms at Wari Bateshwar in the Narsingdi area of the Meghna estuary.
6. *Rajasthan*: Noh, Ahar (black-and-red ware deposit and Painted Grey Ware at Noh, black-and-red ware deposit at Ahar). The early historical sites which have been considered are Bairat, Sambhar, Rairh and Nagari.
7. *Gujarat*: the earliest data are early historical: Prabhas Patan, Dhatwa, Timbarva, Nagara, etc. Devnimori (later history).
8. *Malwa*: Nagda, Eran, Ujjain, Besnagar (the periods IIA at Nagda and Eran and the periods I at Ujjain and Besnagar). The early historical levels which have been considered are Besnagar, Maheshwar-Navdatoli, Nagda and Tripuri.
9. *Vidarbha*: the megalithic sites discussed are Takalghat, Khapa and Gangpur. Also, references to Naikund and Mahurjhari. Among the early historic sites, Paunar is mentioned.
10. *The Deccan*: Prakash (Period II), Nevasa (Periods IV and V), Nasik (Period IIA).

11. *South Indian megaliths*: general discussion according to the major tool-types; also specific reference to Adichanallur.
12. *Orissa*: Sisupalgarh (early historic level).

The number of ancient iron objects on which technical studies have been carried out must be considered lamentably inadequate. The basic list of such objects is as follows:

1. Delhi pillar (fourth century A.D.—wrought iron).
2. Dhar pillar (possibly fourteenth century A.D.—wrought iron).
3. Konarak beam (thirteenth century A.D.—mild steel).
4. Iron from below the Heliodorus pillar at Besnagar (second century B.C.—steel)..
5. Taxila samples (mostly from Sirkap): double-edged sword (high carbon steel), fragmentary sword (high carbon steel), dagger (medium to low carbon steel), dagger (wrought iron), carpenter's adze (high carbon steel), axe (wrought iron), chisel (wrought iron), knife (wrought iron), arrowhead (medium or low carbon steel), double-edged spearhead (wrought iron).
6. Kausambi samples: arrowhead (100 B.C.), arrowhead (300–200 B.C.), arrowhead (100 B.C.–A.D. 500), arrowhead (480 A.D.), arrowhead (395–325 B.C.), iron piece (200 B.C.). All these specimens contain low to medium carbon.
7. Rajghat samples: blade, arrowhead, nail, piece of a rod, a fragmentary piece, nail. All these pieces are dated 600–400 B.C. and the last-mentioned nail specimen shows a high percentage of carbon (1.4 per cent).
8. Kangra sample: knife (c.200 B.C.—evidence of quenching).
9. Atranjikhera samples: 4 samples from the Painted Grey Ware deposit—all wrought iron with evidence of carburisation.
10. Barudih sample: sickle—low carbon steel.
11. Prakash sample: a shaft-hole axe (c.200 B.C.—wrought iron with evidence of carburisation).
12. Mahurjhari sample: axe (·9 per cent carbon).
13. Takalghat—Khapa sample: spear ('may be a variety of steel').
14. Raigir sample: megalithic trident (·70–·85 per cent carbon).
15. Tadakanahalli sample: early context of south Indian megaliths—wrought iron with evidence of carburisation.
16. Naikund sample: megalithic iron—medium carbon steel.
17. Dhatwa sample: early centuries A.D.—manufacturing process of a hoe.

It is important to note that the evidence of steel is quite early. The Barudih iron sickle should date from 1000 + B.C. and it is found to be made of low carbon steel. The megalithic iron objects from Vidarbha and the areas further south seem to contain a fair number of steel specimens. In this context it is interesting to recollect that the iron objects from the central Indian megaliths have been known to contain steel specimens for a long time. When the objects found in G. G. Pearse's excavation of a stone circle near Kamptee in 1869 were given to the British Museum they were found to be of steel and assigned a date of about 1500 B.C. We have pointed out (chapter 1) that this antiquity of Indian steel has been mentioned in various histories of technology.

TEXTUAL REFERENCES

On the basis of our review of the literary sources (chapter 4) it should be possible to make the following generalizations regarding the nature of early Indian literary data on iron:

1. Though it is not possible to determine the exact point of the beginning of iron on the basis of literary data, the evidence is clear enough to suggest that iron was associated with agriculture, at least in the upper Gangetic valley and the Indo-Gangetic divide, around 700 B.C. There are also reasonable grounds for pushing the process back to c. 800 B.C. In the middle Gangetic valley there is positive literary evidence that iron was used in agriculture in c.500 B.C., but this does not, of course, mean that there was no earlier use in this region.
2. The meaning of *ayas* in the *Ṛgveda* cannot be satisfactorily determined because there is no clue in the contexts of its occurrence. It was perhaps a generic term which might have suggested not merely copper-bronze but also, occasionally, iron. As a generic term this occurs also in the *YV* and *AV*, both of which were, however, familiar with iron. But from the *ŚB* onwards this was the basic term used to denote iron. The other terms like *śyāma, kṛṣṇāyasa*, and their variants were used, but only sporadically. *Tīkṣṇa* possibly denoted steel. This occurs first in the *Arthaśāstra* but it could have been earlier. The term *asiloha* in the sense of iron or any of its specific varieties occurs only once. The meaning of the term *vṛtta* is uncertain.
3. There is no significant reference to the details of iron metallurgy. There are references, of course, to different types of iron implements like spikes, chains, cauldrons, etc. but they do not add up to anything significant historically. Considering the general character of early Indian literature this, however, is not surprising. The evidence of later texts like *Bṛhatsaṃhitā* and *Rasaratnasamucchaya* seems to suggest the existence of a metallurgical literature which is now lost.

PRE-INDUSTRIAL SMELTING

The basic data we have cited on the pre-industrial iron-smelting tradition may be taken to be representative of what was current in the middle of the nineteenth century and slightly later. Although the relevant writings of both Ball and Mallet were published, respectively, in 1881 and 1890, they were obviously referring to the data studied earlier than that period. Three preliminary points emerge from their lists of the ore-bearing area with pre-industrial smelting.

First, although the distribution of ores and smelting is ubiquitous in the relevant zones, there were several major production centres in each of these zones. In south India, the major production centres in Mallet's list were the following: the Shenkotta taluk in Travancore; five taluks in Malabar, which produced among themselves 475 tons of iron annually; south and north Arcot; Guntur and Masulipatam areas of the Kistna district; the Jaipur zamindari in the Vizagapatam district; and finally, Hyderabad. In central India, the major centres in Mallet's time were Raipur, Narsinghpur (cf. Tendukhera), Jabalpur, Bundelkhand, Lalitpur and Banda. In west India, one may mention Rewa Kantha in Gujarat. In Rajasthan, Alwar and Gurgaon were important areas. In the Chhotanagpur plateau of eastern India, Hazaribagh was the most important centre. In the northern mountains, one may refer to Bannu, Peshawar, Kashmir, the Panjab Himalayas and Kumaun. The iron industry in north-east India was not particularly significant when Mallet compiled his list.

Secondly, by the time Ball and Mallet set down their observations, the indigenous iron industry was losing ground not merely because of competition with English iron but also because of the growing scarcity of fuel, i.e., charcoal. In a number of places (cf. Salem) Mallet makes this point clear.

Thirdly, as far as the production of steel was concerned, it was then produced only in a few areas mainly in the south. The areas which have been mentioned specifically as steel-producing by Mallet are the Vanga-colum village in Tinnevelli, south Arcot, the Bangalore and Nagar divisions of Mysore and Hyderabad. In central India, we first draw attention to Mallet's report in the case of Tendukhera in Narsinghpur that a part of the local iron was 'converted into steel, by a method different from that practised in Madras'. In Jabalpur he referred to iron of 'a hard steely kind used for edged tools'. In Darjeeling there was 'a steely iron suitable for making *kukris* and *bans*'. It is quite possible that the steel-making centres were once far more widespread. However, this point cannot be settled without further research on the historical records.

In our account of the general smelting operation we first outlined the three basic types of furnaces described by Percy. The simplest type was also the most common. The other two types were necessary only for large-scale operations. The third type involved a rather complicated mode of furnace-construction, but as Percy points out, this was generally used to produce better quality iron and natural steel. That there could be variations in the methods of smelting operations has been made clear by H. F. Cleer's description of the three furnaces which he saw in operation in Jamshedpur in February 1963. These furnaces, one may presume, were once current in the Singhbhum area of Bihar, where Jamshedpur is located. The furnaces were given the names of the villages which sent them for the exhibition in Jamshedpur where Cleere saw them: the Kamar Joda, Single Becha and Jiragora furnaces. The Jiragora furnace conforms to the most common type described by Percy but the other two types constitute new additions to the general range of the pre-industrial iron-smelting furnace types in the subcontinent. What is striking about the Kamar Joda and Single Becha furnaces is that both of them are more primitive than the general type of primitive iron-smelting furnaces described by Percy and others in the Indian context. In the Kamar Joda furnace, the temperature did not exceed 1100 degrees centigrade. There was no provision for the removal of molten slag and the metal had to be obtained by collecting the metal particles scattered in the slag matrix. There was provision for the removal of molten slag in the Single Becha furnace. This, however, was constructed at the bottom of a large square pit.

From the records we have cited, it appears that there were considerable inter-regional differences in the details of iron manufacture in different areas. There were variations in the methods of steel manufacture too. The description of the general process by Buchanan and Heyne is well known. In this process pieces of iron were put in crucibles along with some vegetable matter. However, as we have pointed out, David Mushet wrote on the authority of Heath that in certain situations grey cast iron (produced in a furnace about 8 ft. high, using four times the normal quantity of charcoal required) was put in the crucibles as a substitute for the required vegetable matter. Secondly, according to the testimony of Voysey, the size of the crucibles depended on the purpose for which the steel was intended.

Voysey also saw a piece of glass being put in the crucibles to act as a flux. Thirdly, wc have cited Cecil Von Schwarz on the method of producing 'crucible cast steel'. This record is unique in the sense that in the Indian context no other testimony refers to this method by which steel was obtained directly from the ores.

In the section on the regional details, we have cited data from Ferozepur in the Gurgaon area, Kathiawar, miscellaneous places recorded by Buchanan in the south, the Khasi hills, the Orissan uplands and the village of Mirjati near Jamshedpur. These details perhaps give us an insight which no generalized description of pre-industrial iron-smelting can offer. It is indeed interesting to reflect that, as late as 1918, a pre-industrial iron-smelter and blacksmith coult satisfactorily ply his trade within walking distance of the steel city of Jamshedpur. What is equally remarkable is that the iron of the Delhi pillar was manufactured in the same way as the iron of the Mirjati village near Jamshedpur.

CONCLUDING OBSERVATIONS

In this volume we have analysed in detail the currently available sources of the early history of iron in India. We have noted that there are at least two regions in which the beginning of iron may, with a great deal of justification, be put in the second half of the second millennium B.C. The occurrence of iron in the black-and-red ware deposit at Ahar in south-east Rajasthan and in what has been called the pre-Northern-Black-Polished-Ware and post-chalcolithic black-and-red ware phases of Nagda, Eran, Dangwada and Prakash in Malwa and the periphery of Malwa decidedly goes beyond 1000 B.C., and at present the weight of the evidence takes it back to the middle of the second millennium. We believe that the Ahar evidence which comes from five trenches and is distributed in different layers cannot be wished away by merely suggesting that it came from disturbed layers. The issue is clinched by the discovery of iron at Lothal. The beginning of Nagda II was put around 750 B.C. by Banerjee on the basis of his premise that the Malwa culture at Navdatoli came to an end around 800 B.C., whereas the current radiocarbon evidence puts this date around 1400 B.C. The radiocarbon dates from the early iron-bearing level of Eran were either ignored or bracketed with the earlier chalcolithic dates from the site for no other apparent reason than a general unwillingness to face early dates for iron in Malwa. In Karnataka, in south India, the early radiocarbon dates of the level of neolithic-megalithic transition at Hallur are now supported by the thermoluminiscent dates from the relevant context of a neighbouring site. This is an important development and clearly strengthens the case of a pre-1000 B.C. beginning of iron in south India. Both in the upper Gangetic valley and eastern India the use of iron began unmistakably around 1000 B.C. This is a date which has been invoked for the beginning of iron in the north-western region as well.

Our analysis of the literary data does not suggest much regarding the beginning of iron but we have demonstrated that by the time of the *Śatapatha Brāhmaṇa* (c. 800/700 B.C.) iron was a common enough metal to be associated with the ordinary people.

The archaeological picture clearly suggests India to be an independent and early centre of iron technology. That this cannot be explained by any diffusionary postulate was forcefully argued by me in 1976, and nothing has changed in the meanwhile to discard that argument. If anything, the case for an independent beginning of iron in India has been considerably strengthened in recent years. Having pointed out the early association of iron in south-east Rajasthan, Malwa and eastern India with the black-and-red ware (BRW) deposits, Shaffer writes:

> This association of iron artifacts with BRW in the late second millennium B.C. should *not* be interpreted as representing simply an earlier diffusion of iron technology into the subcontinent. BRW pottery, or any similar type ceramic, is unknown in regions west of the Indus valley. This suggests that BRW pottery and associated cultural traits are entirely of an indigenous South Asian origin. Moreover, the nature and context of the iron objects involved are very different from early iron objects found in Southwest Asia. Most BRW artifacts appear to be utilitarian tools (points, chisels, sickles, axes, nails, knives, crow-bars, etc.). Similar utilitarian iron tools are not generally found in Southwest Asia until *ca.* 850 B.C. The Iranian Plateau Iron Period I may date to the second half of the second millennium B.C. if MASCA dates are used. However, most of these objects are associated with burials unlike BRW iron artifacts which are found in general habitation contexts. A significant number of early Iranian iron objects are items of personal adornment (jewelry) and of the remainder (e.g. daggers), it is difficult to determine if they were utilitarian, ceremonial or status-linked objects because of their burial association. Therefore, the context of functional nature of early iron artifacts in Southwest Asia differ significantly from those in the subcontinent. The context, early dates and different functional nature of iron artifacts in the subcontinent,

suggest that iron technology was an indigenous development and net diffused from some Western source.[7]

Although the archaeological evidence is firmly in favour of an indigenous beginning of iron technology in India, one has to explain it in technological terms. Among the Indian scientists, K.T.M. Hegde has cogently analysed the situation.

> First of all, we have noted the evidence of a long tradition of advanced copper technology in the country. This technology will have provided the most essential infrastructure for smelting iron ore, that is, the furnace. Secondly, the earliest levels of the Iron Age sites have yielded only a few small fragmentary unidentifiable iron objects. This evidence suggests that at this stage the Indian iron industry was in an uncertain experimental stage. There is a reason for that.
> The technique required for smelting iron is a little more difficult than that required for smelting copper. Iron melts at a much higher temperature (1534° c) than copper (1083° c). Also affinity of iron to oxygen is much stronger than that of copper to oxygen. Iron ore is associated with more impurities than copper ore. Iron, therefore, requires more critical conditions for its successful smelting. A temperature of 1250° C is necessary in the furnace to achieve separation of the unwanted gangue materials from the smelting charge. To obtain this high temperature the furnace will need a good supply of oxygen. With such a supply of oxygen it is difficult to maintain reducing condition in the furnace. The smelter has to know how to balance these conflicting demands. He has to maintain a strong blast of air through the furnace. To offset its oxidising effect it is necessary to feed the furnace with an excess of fuel at regular intervals, so that the reducing gas, carbon monoxide, produced in the furnace, dominates over carbon dioxide. All such technical prerequisites appear to have been gradually understood by the early Indian iron smelters. We have evidence of their experiments in the form of small quantities of poor quality iron occurring at the earliest levels of the Iron Age sites. It is, therefore, clear that we did not receive the technical know-how of iron smelting and smithery from an outside source.[8]

We believe that the evidence of the Kamar Joda pre-industrial furnace is very significant in this context. Temperature could not rise above 1100° c in this furnace, and thus no 'separation of the unwanted gangue materials from the smelting charge' could be achieved here. But that was no deterrent to the growth of iron technology because, as we have noticed on the basis of the testimony provided by H. F. Cleere,

[7] Shaffer, 1989.
[8] Hegde, 1981.

there was no provision for the separation of the molten slag from the metal and the metal had to be obtained by collecting the metal particles scattered in the slag matrix. This is the most primitive type of iron-smelting furnace recorded in the subcontinent, and the fact that such a method of producing iron continued in India till the early sixties is a likely indication of an independent transition to iron-smelting in India. On the basis of the existence of such a primitive pre-industrial iron-smelting furnace we may even go to the extent of suggesting that the transition to the use of iron could take place, quite independently of each other, in a number of regions of India.

That the use of iron could begin as a byproduct of copper technology was apparently first suggested by J. A. Charles[9] and, on that basis, Merrick Posnansky[10] argued the possibility of iron-working developing independently in Africa in the first quarter of the first millennium B.C. In this context one may draw attention to M.D.N. Sahi's communication to the Annual Conference of the Indian History Congress in 1980. He points out that the iron content of the three samples of the copper metallurgical slag from the chalcolithic Ahar, which have been analysed by Hegde, is 43.89 per cent, 45.32 per cent and 48.26 per cent. Sahi depends on Hegde's observations to suggest that the Ahar coppersmiths could tackle the problem of alienating iron from copper ore and thus must have been conversant with the properties of iron ore. Moreover, Sahi argues, following Hegde, that one of the conditions for the removal of iron from copper is the high temperature of the smelting furnace, over 1200° C. The Ahar chalcolithic furnaces could obviously raise this temperature and in such furnaces iron could be expected to form a semi-fluid mass that could be forged and worked. Sahi concludes that 'iron metallurgy was invented independently as an offshoot of copper metallurgy in India, particularly in the regions of Malwa and Banas cultures'.[11]

In archaeological terms what should be the chronological bracket of the Indian Iron Age? In 1983 we argued that this should comprise the time-span from the beginning of iron in the archaeological sequence of an area to the beginning of the historical period in that area.[12] According to this approach, the chronological bracket of the Iron Age in the upper Gangetic valley is between C. 1100/1000 B.C.,

[9] Charles, 1979.
[10] Posnansky, 1982.
[11] Sahi, 1980.
[12] Chakrabarti, 1983.

when the use of iron began in this area, to c. 700/600 B.C. when the historical period was heralded by the Northern Black Polished Ware.

If one is asked to comment on the further direction of research on the history of iron and steel metallurgy in India, one may point out the following. First, the early literary data are unlikely to throw much new light on either the beginning or the metallurgical details of iron manufacture. However, mediaeval literary sources of various kinds need a detailed study from the metallurgical point of view. The evidence of the Tamil literature and south Indian inscriptions in general is not known to have been analysed for their references to iron and steel. In view of the fact that south India was the home of pre-industrial Indian steel, this task is obviously significant.

Secondly, our discussion has possibly highlighted the great quantity of records on pre-industrial iron-smelting in different areas. As far as practicable these sources need datailed regional studies. They may not add further to our knowledge of the pre-industrial smelting processes but, as far as the organization and economic aspects of pre-industrial iron production are concerned, the full potential of these sources still remains to be tapped. Thirdly, the metallographic study of chronologically and typologically representative samples of slag and artefacts in different areas is possibly the most important task ahead. Our study has possibly brought out the remarkable paucity of work in this field. One of the burning problems is to understand the history of steel production. The real antiquity of wootz, for instance, is unknown. The same is true of the production of cast iron in India.

Fourthly, although iron objects continue to be discovered in archaeological excavations, some attention should possibly be directed at the excavations of specifically iron-smelting and manufacturing sites. There are traces of such sites in different iron-ore-bearing areas of the country and some attempts are necessary to understand them both in terms of their dates and the metallurgical practices. Only two such sites—Dhatva and Naikund—have been excavated, underscoring the need for many more excavations of this type in future.

Another discovery in this direction has also been made in south India, at the megalithic site of Kodumnal in the Periyar district of Tamil Nadu. The megalithic level here has been dated to c. 250 B.C. In his article 'New Light on the Megalithic Culture of the Kongu Region, Tamil Nadu' (*Man and Environment*, vol. 15, no. 1, 1990, pp. 93–102), K. Rajan reports the following: 'A large oval-shaped furnace surrounded by a number of small furnaces was exposed in the bottommost level of the trenches taken on the northern fringe of the habitation. Vitrified crucibles collected here which appear to be *in situ* may suggest that the large furnace was used to stack up a number of crucibles containing iron and carbon to produce steel by prolonged hearing.' This observation is entirely in agreement with the description of steelmaking in south India in Buchanan's and Heyne's records. This is the earliest record of its kind.

References

Abbott, J.,

1843 *Narrative of a Journey from Heraut to Khiva, Moscow and St. Petersburgh*. London.

——

1847a Process of Working the Damascus Blade of Goojrat (Punjab). *Journal of the Asiatic Society of Bengal* (hereafter cited as *JASB*), pp. 417–23.

——

1847b Additional Observations on the Damask Blades of Goojrat. *JASB*, pp. 666–7.

——

1848 On the Manufacture of the Matchlock of Koteli. *JASB*, pp. 217–80.

Agrawal, D. P.,

1982 *The Archaeology of India*. London.

Agrawal, O. P.,

1983 Scientific and Technological Examination of some Objects from Atranjikhera, in R. C. Gaur, *Excavations at Atranjikhera*, pp. 487–98. Delhi.

Agrawala, V. S.,
1953 *India as known to Pāṇini*. Lucknow.

Alcock, L.,
1952 The Dark Age in Northwest India. *Antiquity* 26:93–5.

Allchin, Bridget and F. R.,
1982 *The Rise of Civilization in India and Pakistan*. Cambridge.

Allchin, F. R.,
1954 *The Development of Early Cultures in the Raichur Doab*. Ph.D thesis, London Univeristy.

——
1970 A Piece of Scale Armour from Shaikhan Dheri, Charsada. *Journal of the Royal Asiatic Society* (hereafter cited as *JRAS*), pp. 113–20.

Altekar, A. S. and Mishra, V.,
1959 *Report on Kumrahar Excavations 1951–55*. Patna.

Andersen, D. and Smith, H. (eds.),
1913 *The Suttanipāta*. London.

Anonymous
1831 Ferozepur Iron Works. *Gleanings in Science* 3:327–8.

Ansari, Z. and Mate, M. S.,
1966 *Excavations at Dwarka*. Poona.

Antonini, C. S.,
1969 Swat and Central Asia. *East and West* (hereafter cited as) *EW* 19: 100–15.

——
1973 More about Swat and Central Asia. *EW* 23: 235–44.

Antonini, C. S. and Stacul, G.,
1972 *The Protohistoric Graveyards of Swat (Pakistan)*, 2 vols. Rome.

Apte, V. S.,
1957 *The Practical Sanskrit-English Dictionary, Vol.I.*, Poona.

Athavle, V. T.,
1965 Iron. *Ancient India* 20–21: 122–39.

Aufrecht, T. (ed).,
1861 *Halāyudha's Abhidhānaratnamālā: A Sanskrit Vocabulary*. London.

Baker, W. E.,
1854 Memorandum on the Prospect of Remuneration in Working the Iron Mines of the Raneegunge District. *JASB*, pp. 484–91.

Ball, V.,
1880 *Jungle Life in India*. London.

——
1881 *A Manual of the Geology of India: Part III—The Economic Geology*. London.

Banerjee, M. N.,
1927 On Metals and Metallurgy in Ancient India. *Indian Historical Quarterly* 3: 121–33; 793–802 (hereafter *IHQ*).

——
1929 Iron and Steel in the Ṛgvedic Age. *IHQ* 5: 432–40.

——
1932 A Note on Iron in the Ṛgvedic Age. *IHQ* 8:364–6.

Banerjee, N. R.
1965. *The Iron Age in India*. Delhi.

——
1986 *Nagda*. Delhi.

Bardgett, W. E. and Stanners, J. F.,
1963 The Delhi Pillar—A Study of the Corrosion Aspects. *National Metallurgical Laboratory Journal* (hereafter *NMLJ*), pp. 24–31.

Belaiew, N.,
1918 Damascene Steel. *Journal of the Iron and Steel Institute (JISI)*, pp. 417–37.

Belck, W.,
1907 Die erfinder der eisentechnik. *Zeitschrift fur Ethnologie* (*ZE*) 39: 334–81.

——
1908 Die erfinder der eisentechnik. *ZE* 40: 45–9.

Belck, W.,
1910 Die erfinder der eisentechnik. *ZE* 42: 15–30.

Benfey, T.,
1866 *A Sanskrit-English Dictionary*. London.

Bhandarkar, D. R.,
1913–14 Excavations at Besnagar. *Annual Report, Archaeological Survey of India (ARASI)*, pp. 186–226.

——
1929 *The Archaeological Remains and Excavations at Nagari*. Calcutta.

Bharadwaj, H. C.,
1973 Aspects of Early Iron Technology in India, in D. P. Agrawal and A. Ghosh (eds.), *Radiocarbon and Indian Archoeology (RIA)*, pp. 391–400.

Bilgami, S. A.,
1899 An Account of the Iron Industry in the Territory of the Nizam of Hyderabad. *JISI* 78: 65–82.

Bohtlingk, Otto and Roth, Rudolph
1855 *Sanskrit-Worterbuch*. St. Petersburg.

Bose, A. N.,
1945 *Social and Rural Economy of Northern India, c. 600 B.C. - c.200 A.D.*, 2 vols. Calcutta.

Bréant, J. R.,
1823 Description d'un procédé à l'aide duquel on obtient une espèce d'acier fondu semblable à celui des lames damassées orientales. *Bulletin de la Société d' Encouragement pour l' Industrie Nationale* 22: 222–7.

Brose, H. L.,
1930 Comments, in A. Neuburger, *The Technical Arts and Sciences of the Ancients*. London.

Buchanan, F.,
1807 *A Journey from Madras*. 3 vols. London.

Cappeller, C.,
1891 *A Sanskrit-English Dictionary*. London.

Chakrabarti, Dilip K.,
1973 Beginning of Iron and Social Change in India. *Indian Studies: Past and Present* 14: 329–38.

Chakrabarti, Dilip, K.,
1974 Beginning of Iron in India: Problem Reconsidered, in A. K. Ghosh (ed.), *Perspectives in Palaeoanthropology: D. Sen Festschrift*, pp. 345–56. Calcutta.

——
1976 The Beginning of Iron in India. *Antiquity* 50: 114–24.

——
1977 Distribution of Iron Ores and the Archaeological Evidence of Early Iron in India. *Journal of the Economic and Social History of the Orient* (JESHO) 20: 166–84.

——
1977 Research on Early Indian Iron, 1795–1950. *The Indian Historical Review* 4: 96–105.

——
1979 Iron in Early Indian Literature. *JRAS*, pp. 22–30.

——
1985 The Issues of the Indian Iron Age, in S. B. Deo and K. V. Paddayya (eds.), *Recent Advances in Indian Archaeology*, pp. 74–7. Pune.

Chakrabarti, Dilip K. and Hasan, S. J.,
1982 The Sequence at Bahiri (Chandra Hazrar Danga), District Birbhum, West Bengal. *Man and Environment* (*ME*) 6: 111–49.

——
1984 *The Antiquities of Kangra*. Delhi.

Chalmers, R. (ed.),
1899 *The Majjhimanikāya*, vol. III. London.

Charles, J. A.,
1979 From Copper to Iron—The Origin of Metallic Materials. *Journal of Metals* 30: 8–13.

Chattopadhyay, P. K.,
1984 Archaeometallurgical Studies in Indian Subcontinent: A Survey of Metallography of Iron Objects. *Indian Journal of History of Science* 19: 361–5.

Cleere, H. F.,
1963 Primitive Indian Iron-making Furnaces. *The British Steelmaker*, pp. 154–8.

Cowell, E. B.,
1895–1907 *The Jātaka*. Cambridge.

Cracroft, W.,
1832 Smelting of Iron in the Kasya (Khasia) Hills (Assam). *JASB*, pp. 150–1.

Cunningham, A.,
1871 *Archaeological Survey of India Reports* (*ASIR*) I. Simla.

Dani, A. H.,
1965–6 Shaikhan Dheri Excavation. *Ancient Pakistan* 2: 119–20 (iron objects).

—— (ed.),
1967 *Ancient Pakistan* No. 3.

Dar, S. R.,
1970–1 Excavation at Manikiyala-1968. *Pakistan Archaeology* 7: 6–22.

Day, St. J. V.,
1877 *The Prehistoric Use of Iron and Steel with Observations on Certain Matters Ancillary Thereto*. London.

De, S. K. (ed.),
1959 *The Droṇaparvam*. Poona.

Deo, S. B.
1970 *Excavations at Takalghat and Khapa*. Nagpur.

——
1973 *Mahurjhari Excavation 1970–72*. Nagpur.

Deo, S. B. and Dhavalikar, M. K.,
1968 *Paunar Excavations*. Nagpur.

Dikshit, M. G.,
1955 *Tripuri*. Nagpur.

Dunn, J. A.,
1942 *The Economic Geology and Mineral Resources of Bihar Province*. Calcutta.

Eggeling, J.,
1885–1900 *The Śatapatha Brāhmaṇa*, 5 vols. Oxford.

Elwin, V.,
1942 *The Agarias*. Calcutta.

Faraday, M.,
1819 An Analysis of *wootz* or Indian Steel. *Quarterly Journal of Literature, Science and the Arts*. 7: 288–90.

Fausboll, V. (ed.),
1901 *The Jātaka*, 6 vols. London.

——
1881 *The Suttanipāta*. Oxford.

Feer, M. L. (ed.),
1894 *The Samyutta-Nikāya*. London.

Foote, R. B.,
1901 *Catalogue of Prehistoric Objects in the Madras Museum*. Madras.

Forbes, R. J.,
1972 *Studies in Ancient Technology*. Vol. 9. Leiden.

Friend, J. N.,
1926 *Iron in Antiquity*. London.

Friend, J. N. and Thornycroft, W. E.,
1924 Examination of Iron from Konarak. *JISI*, pp. 313–15.

Gaur, R. C.,
1983 *Excavations at Atranjikhera*. Delhi.

Goldstucker, T.,
1856 *A Dictionary of Sanskrit and English*. Berlin.

Gopal, R.,
1959 *India of the Vedic Kalpasūtras*. Delhi.

Gordon, D. H.,
1950 The Early Use of Metals in India and Pakistan. *Journal of the Royal Anthropological Institute*, pp. 55–78.

——
1958 *Prehistoric Background of Indian Culture*. Bombay.

Graves, F. G.,
1912 Further Notes on the Early Use of Iron in India. *JISI*, pp. 187–202.

Griffith, R. T. H.,
1896–7 *The Hymns of the Rigveda*. 2 vols. Benares.

Gururaja Rao, B. K.,
1972 *The Megalithic Culture in South India*. Dharwar.

Hadfield, R.,
1912 Sinhalese Iron and Steel of Ancient Origin. *JISI*, pp. 134–72.

——
1925 Science Analyzes the Iron Pillar of Delhi. *Iron Trade Review*. 26 November.

——
1931 *Faraday and His Metallurgical Researches*. London.

——
1951 Technological examination, in J. Marshall, *Taxila* pp. 526–37, 562–3. London.

Halim, M. A.,
1970–1 Excavations at Saraikola Part I. *PA* 7: 23–89.

Hannay, S. F.,
1857 Notes on the Iron Ore Statistics and Economic Geology of Upper Assam. *JASB*, pp. 330–44.

Harris, H.,
1923 Native Manufacture of Wrought Iron in Small Blast Furnaces in India. *JISI*, pp. 42–62.

Hawkshaw, J.,
1875 Presidential Address, in Report of the British Association for the Advancement of Science. Bristol.

Heath, J. M.,
1832 Salem Iron Works. *JASB*, pp. 253–5.

——
1839 On Indian Iron and Steel. *JRAS*, pp. 390–7.

Hegde, K. T. M.,
1973 Early Stages of Metallurgy in India, in *RIA*, pp. 401–5.

——
1981 Scientific Basis and Technology of Ancient Indian Copper and Iron Metallurgy. *Indian Journal of History of Science* 16: 189–201.

Heyne, B.,
1814 *Tracts, Historical and Statistical, on India*. London.

Horner, I. B.,
1954–9 *The Collection of Middle Length Sayings*. 3 vols. London.

Hunday, A. and Banerjee, S.,
1967 *Geology and Mineral Resources of West Bengal.* Calcutta.

Indian Archaeology—a Review.

Jackson, W.,
1845 Memorandum on the Iron Works of Beerbhum. *JASB*, pp. 754–6.

Jacob, L.,
1843 Report on the Iron of Kattywar, its Comparative Value with the British Metal, the Mines, and Mode of Smelting the Ore. *JRAS*, pp. 98–104.

Jacobi, H.,
1895 *Jaina Sutras*. Oxford.

Jacobson, J.,
1979 Recent Developments in South Asian Prehistory and Protohistory. *Annual Review of Anthropology* 8: 467–502.

Jain, J. C.,
1947 *Life in Ancient India as Depicted in the Jain Canons.* Bombay.

Jarrige, J. F. and Enault, J. P.,
1973 Recent Excacations (French) in Pakistan, in *RIA*, pp. 163–72.

Javaji, T. (ed.),
1909 *Manusmriti*. Bombay.

Jettmar, K.,
1967 An Iron Cheek-piece of a Snaffle Found at Timargarha. *Ancient Pakistan* 3: 203–9.

Joshi, A. P.,
1973 Analysis of Copper and Iron Objects, in S. B. Deo, *Mahurjhari Excavation 1970–72*, p. 77. Nagpur.

Kangle, K. P.,
1963 *The Kauṭilyiya Arthaśāstra*, Part II. Bombay.

Katre, S. M.,
1968 *Dictionary of Pāṇini,* Part I. Poona.

Keith, A. B.,
1914 *The Veda of the Black Yajus School, Entitled Taittiriya Samhita*, 2 vols. Cambridge, Mass.

Keith, A. B.,
1920 *Rigveda Brahmanas, the Aitareya and Kauṣitaki.* Cambridge, Mass.

Kielhorn, F. (ed.),
1962 *The Vyākaraṇa-Mahābhāṣya of Patanjali.* Bombay.

Kosambi, D. D.,
1963 The Beginning of the Iron Age in India. *JESHO* 6: 309–18.

——
1965 *The Culture and Civilization of India in Historical Outline.* London.

Krishnan, M. S.,
1951 *Mineral Resources of Madras.* Calcutta.

——
1954 *Iron-ore, Iron and Steel.* Calcutta.

Krishnan, M. S. and Aiyengar, N.K.N.,
1954 *The Iron Ore Deposits of Parts of Salem and Trichinopoly Districts.* Calcutta.

Lahiri, A. K., Banerjee, T. and Nijhawan, B. R.,
1963 Some Observations on Corrosion-resistance of Ancient Delhi Iron Pillar and Present-time Adivasi Iron Made by Primitive Methods. *NMLJ*, pp. 45–54.

Lal, B. B.,
1949 Sisupalgarh 1948. *Ancient India* 5: 62–105.

——
1954–5 Excavations at Hastinapur. *Ancient India* 10–11: 5–151.

Leshnik, L. S.,
1974 *South Indian 'Megalithic' Burials.* Weisbaden.

Lester, I. E.,
1912 Presidential Address, 1911: Indian Iron. (*PSISI*) *Proceedings of the Staffordshire Iron and Steel Institute* 27: 2–20.

Lowe, Thelma,
1989. *Archaeological Field-survey in Northern Andhra Pradesh—the Identification and Distribution of Iron-smelting and Crucible Steel Production Sites.* Mimeographed report.

Macdonell, A.,
1893 *A Sanskrit-English Dictionary*. London.

Mallet, F. R.,
1890 Iron, in George Watt, *A Dictionary of the Economic Products of India*. London, pp. 439–513.

Marcadieu, M.,
1855 Report on the Kooloo Iron Mines and Portion of the Mannikurn Valley. *JASB*, pp. 191–202.

Marshall, J.,
1951 *Taxila*, 3 vols. Cambridge.

Max Muller, F.,
1965 (reprint). *Ṛgveda Saṁhitā*, 2 vols. Varanasi.

——
1881 *The Dhammapada*. Oxford.

McCrindle, J. W.,
1879 *The Commerce and Navigation of the Erythraean Sea*. London.

——
1877 *Ancient India as Described by Megasthenes and Arrian*. London.

——
1882 *Ancient India as Described by Ktesias the Knidian*. London.

McWilliam, A.,
1920 Indian Iron-making at Mirjati, Chhotanagpur. *JISI*, pp. 159–70.

Mehta, R. N.,
1955 *Excavations at Timbarva*. Baroda.

——
1968 *Excavations at Nagara*. Baroda.

Mehta, R. N. and Chowdhary, S. N.,
1966 *Excavation at Devnimori*. Baroda.

Minayeff, Prof. (ed.),
1888 *Pettavatthu*. London.

Mitra, R. L. (ed.),
1870 *The Taittīriya Brāhmaṇa of the Black Yajurveda*. Calcutta.

Morris, R. and Hardy, E. (eds.),
1885–1900. *The Anguttara-Nikāya*, Parts 1–5. London.

Mughal, R.,
1967 Excavations at Tulamba, W. Pakistan. *PA* 4: 11–152.

Mukherjee, B.,
1926 *Rasajālanidhi or Ocean of Indian Chemistry and Alchemy*, vol. II. Calcutta.

Munshi, K. N. and Sarin, R.,
1970 Analysis of Copper and Iron Objects from Takalghat and Khapa, in S. B. Deo, *Excavations at Takalghat and Khapa (1968–69)*, pp. 78–9. Nagpur.

Mushet, D.,
1805 Experiments on *wootz*. *Philosophical Transactions* 95: 163–75.

——
1840 *Papers on Iron and Steel, Practical and Experimental*. London.

Neuburger, A.,
1930 *The Technical Arts and Sciences of the Ancients*. London.

Newbold, T. H.,
1846 Summary of the Geology of Southern India. *JRAS*, pp. 138–71.

Norman, K. R.,
1969 *The Elders' Verses, I. Theragāthā*. London.

——
1971 *The Elders' Verses, II. Therigāthā*. London.

Oldenberg, H. (ed.),
1879 *The Vinaya Pitakam*, vol. I. London.

——
1880 *The Vinaya Pitakam*. vol. II. London.

Oldenberg, H. and Pischell, R. (eds.),
1966 *The Thera-and Theri-Gathā*. London.

Oldham, C. (ed.),
1930 *Journal of Francis Buchanan Kept during the Survey of the District of Bhagalpur in 1810–1811*. Patna.

Parab, K. P.(ed.),
1902 *The Rāmayaṇa of Vālmīki*. Bombay.

Pearse, G. G.,
1869 Notes on the Excavation of a Stone Circle near Kamptee, Central Province of India. *Journal of the Ethnological Society of London*, 1: 207–17.

Pearson, G.,
1795 Experiments and Observations to Investigate the Nature of a Kind of Steel, Manufactured at Bombay, and there called Wootz......*Philosophical Transactions* 85: 322–46.

Percy, J.,
1864 *Metallurgy: Iron and Steel*. London.

——
1886 Presidential address. *JISI*, pp. 8–31.

Piddington, H.,
1855 Memorandum on the Kunkurs of Burdwan as a Flux for Smelting the Iron Ores, and on some Smelting of Iron Ores by Mr. Taylor of that District. *JASB*, pp. 212–15.

Pleiner, R.,
1971 The Problem of the Beginning of the Iron Age in India. *Acta Praehistorica et Archaeologica* 2: 5–36.

Posnansky, M.,
1982 African Archaeology Comes of Age. *World Archaeology* 13: 345–58.

Possehl, G. L.,
1988a Radiocarbon Dates from South Asia. *ME* 12: 169–96.

——
1988b The Coming of the South Asian Iron Age, an Indigenous Technological Process. Mimeographed paper.

Prakash, S. and Singh, R.,
1968 *Coinage in Ancient India*. Delhi.

Punyavijayji, Muni Shri (ed.),
1957 *Aṅgavijjā*. Banaras.

Puri, B. N.,
1968 *India in the Time of Patanjali*. Bombay.

Radhakrishnan, S.,
1953 *The Principal Upaniṣads*. London.

Rahman, A.,
1968–9 Excavation at Damkot. *Ancient Pakistan* 4: 103–250.

Ray, P. C.,
1902 *A History of Hindu Chemistry*, vol. I. Calcutta.

Ray, P. (ed.),
1956 *History of Chemistry in Ancient and Mediaeval India, Incorporating the History of Hindu Chemistry by Acharya Prafulla Chandra Ray*. Calcutta.

Rea, A.,
1902–3 Prehistoric Antiquities in Tinnevelly. *ARASI*, pp. 111–40.

Rhys Davids, T. W. and Oldenberg, H.,
1881–5 *Vinaya Texts*. Oxford.

Rhys Davids, T. W.,
1890–4 *The Questions of King Milinda*. Oxford.

Roer, E. and Cowell, E. B. (eds.),
1860–99 *The Samhitā of the Black Yajurveda*. 6 vols. Calcutta.

Rolfe, J. C.,
1946 *Quintus Curtius*, 2 vols. London.

Roscoe, H. E. and Schorlemmer, C.,
1897 *A Treatise on Chemistry*, vol. II. London.
1831 Account of the Process of Making Iron at Amdeah near Sambalpur (Orissa). *Gleanings in Science* 3: 330.

Roy, B. C.,
1959 *The Economic Geology and Mineral Resources of Rajasthan and Ajmer*. Calcutta.

Roy, T. N.,
1983 *The Ganges Civilization*. Delhi.

Roy Choudhuri, M. K.,
1955 *Economic Geology and Mineral Resources of Madhya Bharat*. Calcutta.

Rydh, H.,
1959 *Rangmahal*. Lund.

Sahi, M. D. N.,
1979 Iron at Ahar, in D. P. Agrawal and Dilip K Chakrabarti (eds.), pp. 365–6. Delhi.

——
1980 Origin of Iron Metallurgy in India. *Proceedings of the Indian History Congress*. Delhi.

Sahlins, A.,
1913 Potentialities of India as an Iron-producing Country, and Description of the Tata Iron and Steel Works. *PSISI* 28: 50–72.

Sahni, D. R.,
n.d.(a). *Archaeological Remains and Excavations at Bairat*. Jaipur.

——
n.d.(b). *Archaeological Remains and Excavations at Sambhar*. Jaipur.

Sankalia, H. D., Deo, S. B., Ansari, Z. and Ehrhardt, S.,
1960 *From History to Prehistory at Nevasa*. Poona.

Sankalia, H. D. and Deo, S. B.,
1955 *Report on the Excavations at Nasik and Jorwe*. Poona.

Sankalia, H. D., Subbarao, B. and Deo, S. B.,
1958 *The Excavations at Maheshwar and Navdatoli 1952–53*. Baroda.

Sankalia, H. D., Deo, S. B. and Ansari, Z.,
1969 *Excavations at Ahar*. Poona.

Sastrigal, S. C.(ed.),
1912 *Aṣṭādhyāyī sūtrapāṭha*. Trichinopoly.

Schoff, W. H.,
1915 The Eastern Iron Trade of the Roman Empire. *JAOS* 35: 224–39.

——
1974 (reprint). *The Periplus of the Erythraean Sea*. Delhi.

Schroeder, L. Von (ed.),
1881–3 *Maitrāyaṇī Saṁhitā*, 2 vols. Leipzig.

——
1900 *Kāṭhakaṁ*. Leipzig.

Schwarz, C. Von.
1899 Comments on Bilgami's paper. *JISI* 78: 89–99.

——
1901 Ueber die eisen-und stahlindustrie ostindiens. *Stahl und Eisen*, pp. 208–11, 277–83, 337–41, 391–9.

Shaffer, J. G.,
1978 Bronze Age Iron from Afghanistan: Its Implications for South Asian Protohistory. Mimeographed Paper.

——
1989 Mathura: A Protohistoric Perspective in D. M. Srinivasan (ed.), *Mathura, the Cultural Heritage*, pp. 171–80. Delhi.

Shamasastry, B. (ed.),
1919 *Arthaśāstra of Kautilya*. Mysore.

Shamasastry, R.,
1929 *Kauṭilya's Arthaśāstra*. Mysore.

Sharif, M.,
1969 Excavation at Bhir Mound, Taxila. *PA* 6: 6–99.

Sharma, G. R.,
1960 *The Excavations at Kausambi 1957–59*. Allahabad.

Sharma, Y. D.,
1953 Exploration of Historical Sites. *Ancient India* 9: 116–69.

Shastree, B. G.,
1844 Notes on the Iron Ore in the Vicinity of Malvan. *Journal of the Bombay Branch of the Royal Asiatic Society*, pp. 435–7.

Singh, S. D.,
1965 *Ancient Indian Warfare*. Leiden.

Sircar, D. C.,
1966 *Indian Epigraphical Glossary*. Delhi.

Sisco, A. G.,
1956 *Reaumur's Memoirs on Steel and Iron*. Chicago.

Smith, C. S. (ed.),
1968 *Sources for the History of the Science of Steel 1532–1786*. Cambridge, Mass.

Smith, V.
1897 The Iron Pillar at Dhar. *JRAS* 29: 143–6.

Smith, V.
1912 Comments on Hadfield. *JISI*, pp. 183–4.

Stacul, G.,
1966a Preliminary Report on the Pre-Buddhist Necropolises in Swat (W. Pakistan). *EW* 16: 37–8.

—— 1966b Notes on the Discovery of a Necropolis near Kherai in the Gorband Valley. *EW* 16: 261–74.

—— 1967a Explorations in a Rockshelter near Ghaligai (Swat, W. Pakistan). *EW* 17: 185–219.

—— 1967b Discovery of Four Pre-Buddhist Cemeteries near Pacha in Buner. *EW* 17: 220–32.

—— 1969a Excavation near Ghaligai (1968) and Chronological Sequence of Protohistorical Cultures in the Swat valley. *EW* 19: 44–91.

—— 1969b Discovery of Protohistoric Cemeteries in the Chitral Valley (W. Pakistan). *EW* 19: 92–9.

—— 1970a The Grey Pottery in the Swat Valley and the Indo-Iranian Connections (1300–300 B.C.). *EW* 20: 92–102.

—— 1970b An Archaeological Survey near Kalam (Swat Kohistan). *EW* 20: 87–91.

—— 1971 Cremation Graves in Northwest Pakistan and their Eurasian Connections: Remarks and Hypotheses. *EW* 21: 9–19.

—— 1973a Ochre-coloured and Grey-burnished Wares in Northwest Indo-Pakistan (c.1800–1300 B.C.). *EW* 23: 79–88.

—— 1973b A Decorated Vase from Gogdara (Swat Pakistan). *EW* 23: 245–8.

Stein, A.,
1929 *An Archaeological Tour in Waziristan and Northern Baluchistan*. Delhi.

——
1931 *An Archaeological Tour in Gedrosia*. Delhi.

Stede, W. (ed.),
1918 *Niddesa II Cullaniddesa*. London.

Steinthal, P. (ed.),
1885 *Udānam*. London.

Stodart, J.,
1818 A Brief Account of *Wootz* or Indian Steel, Showing its Fitness for Making Surgical Instruments and Other Articles of Fine Cutlery. *Asiatic Journal* 5: 570–1.

Subrahmanyam, B. R.,
1964 Appearance and Spread of Iron in India—An Appraisal of Archaeological Data. *Journal of the Oriental Institute*, Baroda 13: 349–59.

Swank, J. M.,
1892 *History of the Manufacture of Iron in All Ages, and Particularly in the United States from Colonial Times to 1891*. Philadelphia.

Thapar, B. K.,
1957 Maski 1954. *Ancient India* 13: 4–142.

——
1964–5 Prakash 1955. *Ancient India* 20–1: 5–167.

Thatte, C. S.(ed.),
1877 *Amarakoṣa*. Bombay.

Thera, S. (ed.),
1914 *The Dhammapada*. London.

Tilak Maharashtra University.
1933 *Ṛgveda Saṁhitā*. Poona.

Torrens, H.,
1851 Specimen of Iron from the Dhunakar Hills, Birbhum *JASB*, pp. 77–8.

Touche, La,
1917 *A Bibliography of Indian Geology and Physical Geography*. Calcutta.

Trenckner, V. (ed.),
1880 *The Milindapañho*. London.

——
1888 *The Majjhimanikāya*, vol. I. London.

Tripathi, V.,
1973 Introduction of Iron in India, in *RIA*, pp. 272–8. Bombay.

Tucci, G.
1958 Preliminary Report on an Archaeological Survey in Swat. *EW* 9: 279ff.

——
1958 The Tombs of the Asvakayana-Assakenoi. *EW* 14: 27–8.

Turner, R. L.,
1966 *A Comparative Dictionary of the Indo-Aryan Languages*. London.

Turner, T.
1893 Production of Wrought Iron in Small Blast Furnaces of India. *JISI* 44: 162–81.

——
1908 *The Metallurgy of Iron*. London.

Vishvavandhu (ed.),
1961–4 *Atharvaveda*, Parts I–IV. Hoshiarpur.

Voysey, H.,
1832 Description of the Native Manufacture of Steel in Southern India. *JASB* 1: 245–7.

Warmington, E. H.,
1928 *The Commerce between the Roman Empire and India*. Cambridge.

Watson, E. R.,
1907 *A Monograph on Iron and Steel Work in the Province of Bengal*. Calcutta.

Watt, G.,
1890 *A Dictionary of the Economic Products of India*. vol. IV. London.

Weber, A. (ed.),
1852 *The Vājasaneyī-Saṁhitā*. Berlin.

Weber, A. (ed.),
1855. *The Śatapatha Brāhmaṇa*. Berlin.

Wheeler, R. E. M.,
1947–8 Brahmagiri and Chandravalli 1947: *Ancient India* 4: 180–310.

——
1959 *Early India and Pakistan*. London.

——
1962 *Charsada*. London.

Whitney, W. D.,
1905 *Atharvaveda Saṃhitā*. 2 vols. Cambridge, Mass.

Wilkinson, H.,
1836 On the Cause of the External Pattern, or Watering of the Damascus Sword-blades. *JRAS* 4: 187–93.

——
1839 On Iron. *JRAS* 7: 383–9.

Williams, M.,
1872 *A Sanskrit-English Dictionary*. Oxford.

Wilson, H. H.,
1850–88 *Rigveda Sanhitā*. 6 vols. London.

Windisch, E. (ed.),
1889 *Itivuttaka*. London.

Winternitz, M.,
1933 *History of Indian Literature*. Calcutta.

Woodward, F. L.,
1927 The Book of the Kindred Sayings. vol. IV. London.

——
1948 *The Minor Anthologies of the Pali Canon*. London.

Woodward, F. L. and Hare, E. M.,
1932–6 *The Book of the Gradual Sayings*. 5 vols. London.

Wynne, F. H.,
1904 Native Methods of Smelting and Manufacturing Iron in Jabalpur, Central Provinces. *JISI*, pp. 578ff.

Yule, H.,
1892 Notes on the Iron of the Kasia Hills. *JASB*, pp. 853–7.

Yule, H. and Burnell, A. C.,
1903 *Hobson-Jobson*. London.

Index